T0135227

Springer Theses

Recognizing Outstanding Ph.D. Research

Aims and Scope

The series "Springer Theses" brings together a selection of the very best Ph.D. theses from around the world and across the physical sciences. Nominated and endorsed by two recognized specialists, each published volume has been selected for its scientific excellence and the high impact of its contents for the pertinent field of research. For greater accessibility to non-specialists, the published versions include an extended introduction, as well as a foreword by the student's supervisor explaining the special relevance of the work for the field. As a whole, the series will provide a valuable resource both for newcomers to the research fields described, and for other scientists seeking detailed background information on special questions. Finally, it provides an accredited documentation of the valuable contributions made by today's younger generation of scientists.

Theses are accepted into the series by invited nomination only and must fulfill all of the following criteria

- They must be written in good English.
- The topic should fall within the confines of Chemistry, Physics, Earth Sciences, Engineering and related interdisciplinary fields such as Materials, Nanoscience, Chemical Engineering, Complex Systems and Biophysics.
- The work reported in the thesis must represent a significant scientific advance.
- If the thesis includes previously published material, permission to reproduce this must be gained from the respective copyright holder.
- They must have been examined and passed during the 12 months prior to nomination.
- Each thesis should include a foreword by the supervisor outlining the significance of its content.
- The theses should have a clearly defined structure including an introduction accessible to scientists not expert in that particular field.

More information about this series at http://www.springer.com/series/8790

Author
Dr. Thomas Owen James
Department of Physics
Imperial College London
London, UK

Supervisor
Prof. Geoffrey Hall
Department of Physics
Imperial College London
London, UK

ISSN 2190-5053   ISSN 2190-5061   (electronic)
Springer Theses
ISBN 978-3-030-31936-6   ISBN 978-3-030-31934-2   (eBook)
https://doi.org/10.1007/978-3-030-31934-2

This Springer imprint is published by the registered company Springer Nature Switzerland AG
The registered company address is: Gewerbestrasse 11, 6330 Cham, Switzerland

Thomas Owen James

A Hardware Track-Trigger for CMS

at the High Luminosity LHC

Doctoral Thesis accepted by
Imperial College London, London, UK

 Springer

For Grandma, who inspired a curiosity and love for all things science and the natural world. I hope that this would have made you proud.

Supervisor's Foreword

The background to Tom James' thesis originates in the requirement to upgrade the CMS experiment at the CERN Large Hadron Collider to extend the studies of physics in the TeV energy range and continue searches for new phenomena for another couple of decades. To make this possible, large parts of the original experiment must be replaced, since they have been exposed to extremely high particle fluxes and will have been damaged by irradiation. At the same time, to successfully study rare processes, including the properties of the Higgs boson first observed in 2012, it will be necessary to acquire even larger event samples and increase the statistical sensitivity of the experiment. This will be made possible by increasing the area and granularity of some of the detectors, especially those closest to the colliding beams, which measure the trajectories of the particles emerging from the collisions. However, even with the fastest and most sensitive electronics available today, it is impossible to capture and store all the data from CMS so it is essential to select events of potential interest using a 'trigger', in which certain characteristics, such as the presence of a very high transverse momentum lepton, signal the possibility of an event of more than usual interest. The electronic trigger for CMS has evolved considerably from its original conception, but still mainly relies on signals from the muon detectors and electromagnetic and Hadron calorimeters. In the future, these signals alone are insufficiently selective to reduce the trigger rate to levels with which CMS can cope. The only way which has been found to improve on this is to exploit information from the tracking detectors in the experiment, which has hitherto not been used. This is immensely challenging, because the number of tracking elements is huge; for triggering purposes, about 13,000 modules comprising about 214 million sensor elements must be used. Studies have demonstrated that the tracking system must provide essentially all fully reconstructed trajectories for particles with transverse momentum above 2 GeV/c and these must be available for the trigger within a few microseconds. This is unprecedented, and especially difficult in a Hadron collider environment which generates multiple events per beam crossing with a very large number of outgoing charged particles from each interaction. Tom's thesis explains how this problem has essentially been solved, even though the final implementation of a system to achieve it has yet to be built, and his, important, role in demonstrating how this will be done. It builds on work carried out, largely at Imperial College, to develop digital electronic hardware to process data from the CMS electromagnetic calorimeter for the Level-1 trigger. Over about twenty years, our group has developed a high level of expertise in programmable digital electronics, based on devices known as FPGAs, which has led to a series of processing boards, which are highly flexible and adaptable to many different problems. Most recently, new concepts were also proposed by our group to deliver a more powerful and flexible trigger for CMS using a time-multiplexing method, which offers many advantages for trigger systems. We decided to attack the track reconstruction problem by applying this new concept. Tom joined the effort in the final year of his undergraduate degree, working initially on software simulations of the tracking problem in CMS. When he started his Ph.D. work the following year, he was well prepared to tackle the full challenge of designing a system and demonstrating how track reconstruction could be implemented using current technology. The subject of Tom's Ph.D., therefore, evolved into a study of how a track-finder could work in CMS to provide the necessary reconstructed high transverse momentum tracks to the L1 trigger with high efficiency, within the available latency. We were able to build a demonstrator system to prove the concept would work by using existing FPGA boards, which had been designed originally for the calorimeter trigger. As the system design, and processing algorithms, evolved we were able to see how it could be made to work and build software and other processing infrastructure to test the idea and, importantly, to be able to compare the results from the FPGA processors with those from realistically simulated CMS events. During his Ph.D., Tom was based in CERN for about 18 months, with the task of implementing our track-finder demonstrator system, collaborating with a few other Imperial staff based there and in London. The success of the demonstrator considerably exceeded what most of us thought was likely initially, and owed much to Tom's efforts. He pushed himself very hard and was dedicated to it, working very long hours to solve any problem he encountered, invariably successfully. Tom demonstrated a remarkable flair for instrumentation work and gained much expertise in advanced software, electronic hardware and the rather complex tools and firmware required to operate modern programmable electronics based on FPGAs, as well as hands-on experience in computer and detector readout systems for real-time data acquisition. His thesis will be a reference for newcomers to the track-finder work for a long time to come. In his thesis, Tom acknowledges many others who contributed to this big task, but I would like to specially emphasise the important long-standing support we have received from our local funding agency, the UK Science and Technology Funding Council.

London, UK   Prof. Geoffrey Hall
November 2019

Abstract

The Compact Muon Solenoid (CMS) experiment at the Large Hadron Collider (LHC) is designed to study a wide range of high-energy physics phenomena. It employs a large all-silicon tracker within a 3.8 T magnetic solenoid, which allows precise measurements of transverse momentum (p_T) and vertex position.

This tracking detector will be upgraded to coincide with the installation of the high-luminosity LHC, which will provide up to about $10^{35}/cm^2/s$ to cms, or 200 collisions per 25 ns bunch crossing. This new tracker must maintain the nominal physics performance in this more challenging environment. Novel tracking modules that utilise closely spaced silicon sensors to discriminate on track p_T have been developed that would only allow the readout of hits compatible with $p_T > 2$–3 GeV tracks to off-detector trigger electronics. This would allow the use of tracking information at the Level-1 trigger of the experiment, a requirement to keep the Level-1 triggering rate below the 750 kHz target, while maintaining physics sensitivity.

This thesis presents a concept for an all field-programmable gate array (FPGA)-based track-finder using a fully time-multiplexed architecture. A hardware demonstrator has been assembled to prove the feasibility and capability of such a system. The track-finding demonstrator uses a projective binning algorithm called a Hough transform to form track-candidates, which are then cleaned and fitted by a combinatorial Kalman filter. Both of these algorithms are implemented in FPGA firmware. This demonstrator system, composed of eight Master Processor Virtex-7 (MP7) processing boards, is able to successfully find tracks in one-eighth of the tracker solid angle at a time, within the expected 4 µs latency constraint. The performance for a variety of physics scenarios is studied, as well as the proposed scaling of the demonstrator to the final system and new technologies.

Acknowledgements

I would like to thank the following people for their support over the last three and a half years:

Geoff Hall
I could not have asked for a better Ph.D. supervisor. Thank you for your seemingly unlimited support and kindness.

Mark Pesaresi
For teaching me almost everything I know in this field. Throughout the enjoyable times, and the difficult times, you have been my mentor, role model and friend. Thank you.

Greg Iles
For never hesitating to offer a lift back to St. Genis after a long day's work; or a coffee in Building 14 when the going gets tough.

Emilija
For unbounded depths of love and support.

Nichola, Alan and Aidan James
For never doubting me. For pushing me to achieve my goals, and supporting me when needed.

In no particular order, I would also like to thank the following people for their intellectual, emotional and practical support throughout the various stages of my Ph.D.: Georg Auzinger, Johan Borg, Erik Butz, Jonathan Fulcher, Marco Garattini, Vito Palladino, Mark Raymond, Andrew Rose, Kirika Uchida and many, many more. Thank you to all, and I look forward to working with you all in the future.

I gratefully thank the Science & Technology Facilities Council (STFC) for funding this research project. Thank you also to the Worshipful Company of Scientific Instrument Makers, for their support via the Postgraduate Scholars Award.

Contents

Acronyms

ADC	Analog-to-digital converter
AIDA	Advanced European Infrastructures for Detectors at Accelerators
ALICE	A Large Ion Collider Experiment
AM	Associative Memory
AM06	Associative Memory 06
AMC	Advanced Mezzanine Card
AMC13	Advanced Mezzanine Card 13
ASIC	Application-specific integrated circuit
ATCA	Advanced Telecommunications Computing Architecture
ATLAS	A Toroidal LHC Apparatus
BDT	Boosted decision tree
BRAM	Block random access memory
BX	Bunch crossing
CBC	CMS binary chip
CERN	The European Organization for Nuclear Research
CIC	Concentrator integrated circuit
CMOS	Complementary metal-oxide semiconductor
CMS	Compact Muon Solenoid
CMSSW	CMS software
CPLD	Complex programmable logic device
CPU	Central processing unit
CSC	Cathode strip chambers
DAQ	Data Acquisition
DDR	Double data rate
DQM	Data quality monitoring
DR	Duplicate removal
DSP	Digital signal processor
DT	Drift tubes
DTC	Data, trigger and control
DTH	DAQ and Timing Hub

ECAL	Electromagnetic calorimeter
ES	Electromagnetic preshower
EU	European Union
FC7	FPGA Mezzanine Card Carrier Kintex-7
FF	Flip flop
FIFO	First in, first out
FILO	First in, last out
FNAL	Fermi National Accelerator Laboratory
FPGA	Field-programmable gate array
FTK	Fast TracKer
GBT	Gigabit transceiver
GLIB	Gigabit Link Interface Board
HB	HCAL barrel calorimeter
HCAL	Hadronic calorimeter
HDL	Hardware description language
HE	HCAL endcap calorimeter
HEP	High-energy physics
HF	HCAL forward calorimeter
HGCal	High-granularity calorimeter
HLHLS	High-level hardware description language
HL-LHC	High-luminosity Large Hadron Collider
HLS	High-level synthesis
HLT	High-level trigger
HT	Hough transform
HTP	Hough transform preprocessor
ILA	Integrated logic analyser
IP	Interaction point
IPbus	Internet protocol bus
IPMI	Intelligent Platform Management Interface
IT	Inner tracker
JTAG	Joint Test Action Group
KF	Kalman filter
KU-115	Xilinx Kintex Ultrascale 115
L1	Level-1
LHC	Large Hadron Collider
LHCb	Large Hadron Collider beauty experiment
LIFO	Last in, first out
LILO	Last in, last out
LpGBT	Low-power gigabit transceiver
LS3	Long Shutdown 3
LUT	Lookup tables
MaPSA	Macro-pixel sub-assembly
MCH	MicroTCA Carrier Hub
MGT	Multi gigabit transceiver
MicroGT	Micro Global Trigger

MicroHAL	Micro Hardware Access Library
microSD card	Micro Secure Digital Card
MicroTCA	Micro Telecommunications Computing Architecture
MP7	Master Processor Virtex-7
MP7-XE	Master Processor Virtex-7 Extended Edition
MPA	Macro-pixel ASIC
MTP	Multi-fibre Termination Push-on
NIM	Nuclear Instrumentation Module
OT	Outer tracker
PC	Personal computer
PCA	Principal component analysis
PCB	Printed circuit board
PCIe	Peripheral component interconnect express
PRM	Pattern Recognition Mezzanine
PS	Proton synchrotron
QCD	Quantum chromodynamics
QDR	Quad data rate
RAM	Random access memory
RCMS	Run control and monitoring system
RMS	Root mean square
RPC	Resistive plate chambers
RTM	Rear transition module
SATA	Serial advanced technology attachment
SERDES	Serialisation/De-serialisation
SEU	Single event upset
SFLR	Seed filter and simple linear regression
SM	Standard model of particle physics
SPS	Super proton synchrotron
SRAM	Static RAM
SSA	Strip sensor ASIC
TFP	Track-finding processor
TIF	Tracker integration facility
TLU	Trigger logic unit
USB	Universal Serial Bus
V7-690	Xilinx Virtex-7 XC7VX690T
VHDL	Very high speed integrated circuit HDL
VL+	Versatile Link PLUS
VU-11P	Xilinx Virtex Ultrascale 11+
VU-9P	Xilinx Virtex Ultrascale 9+
WLCG	Worldwide LHC computing grid
XDAQ	Cross-application Data Acquisition software

Chapter 1
Introduction

1.1 Theory and Motivation

Particle physics is the attempt to describe the universe at its most basic level. The Standard Model of particle physics (SM) is currently the best description of the known universe [1–3]. Limitations and gaps in this theory, however, motivate a continued search for new data to support, or discredit, new and more complete models of Nature. The Large Hadron Collider (LHC) [4], located at CERN, and its future High Luminosity upgrade, the High Luminosity Large Hadron Collider (HL-LHC) [5] are designed for this purpose.

Although the discovery of the predicted Higgs boson at the LHC by both A Toroidal LHC Apparatus (ATLAS) [6, 7] and Compact Muon Solenoid (CMS) [8, 9] experiments further demonstrated the success of the SM, there are a number of outstanding issues:

- The SM is understood to be valid only below an energy scale Λ (which may be as large as the Plank scale), beyond which new physics is needed to describe the universe at higher energy scales.
- Experimental observations of neutrino flavour oscillations imply a non-zero neutrino mass, which is incompatible with the SM hypothesis of a massless neutrino sector [10].
- Cosmological observations suggest that the majority of the mass in the universe is composed of an as yet undiscovered form of matter, which has no viable SM candidate [11, 12].
- Although the SM successfully unifies the weak and electromagnetic forces, it fails to unify these with the strong force, even when extrapolated to high energy scales [13]. In addition, the gravitational force is not incorporated into this model.
- The observed mass of the Higgs boson is predicted to be given by the sum of a bare mass term, and a loop-corrected term which is proportional to the square of the cutoff scale of the theory, Λ^2. The observed Higgs mass of 125 GeV implies that these two terms have a near-perfect cancellation, of up to 36 orders of magnitude. This high degree of *fine-tuning* implies that the theory is not yet complete [13].

© Springer Nature Switzerland AG 2019
T. O. James, *A Hardware Track-Trigger for CMS*, Springer Theses,
https://doi.org/10.1007/978-3-030-31934-2_1

1.2   The Large Hadron Collider

The LHC is a 26.7 km circumference particle accelerator at CERN, Geneva. The LHC is designed to collide protons at up to a centre of mass energy of $\sqrt{s} = 14$ TeV, but is currently running at 13 TeV. It is also able to collide heavy ions such as Pb-Pb, at up 5.02 TeV per colliding nuclear pair [14]. An advantage of colliding protons instead of electrons or positrons is that the increased mass of the proton results in less energy loss via synchrotron emission. As a consequence, however, the high particle flux near the interaction region necessitates a more precise detector with finer granularity, that is also more radiation hard.

The LHC beam must be accelerated in stages. Protons are injected into a linear accelerator known as LINAC2, and then enter the Proton Synchrotron (PS) circular accelerator via a booster. The PS provides a 26 GeV beam (with the required bunch structure) to the Super Proton Synchrotron (SPS) ring, which further accelerates the protons to 450 GeV before filling the LHC ring. The LHC contains two separate beam pipes, each of which accelerates a beam to up to 7 TeV, in opposite directions. Superconducting magnets focus and guide the proton bunches, which cross at four interaction points within the LHC ring. Located at these interaction points are the four main LHC experiments: ATLAS, CMS, Large Hadron Collider beauty experiment (LHCb) [15], and A Large Ion Collider Experiment (ALICE) [16]. While ATLAS and CMS are designed to investigate a wide range of physics at a previously unexplored energy scale, LHCb and ALICE are dedicated to studying CP violation in the B physics sector, and heavy ion physics respectively. During nominal operation, the LHC delivers collisions between bunches of around $N_p = 10^{11}$ protons every 25 ns to its two largest experiments, CMS and ATLAS. Taking an effective cross sectional area of $S_{eff} = 4\pi(16 \times 10^{-4})^2$ cm^2 (not to be confused with the probability of interaction cross section σ_{pp}), and a geometric luminosity reduction factor due to the crossing angle, $F \sim 0.95$, this equates to a design luminosity of

$$L = 40 \times 10^6 \times F \ N_p^2/S_{eff} \sim 1 \times 10^{34}/\,\text{cm}^2/\text{s}. \qquad (1.1)$$

At 7 TeV per proton, the cross section for inelastic proton-proton collisions $\sigma_{in.pp} = 1.1 \times 10^{-25}$ cm^2, giving an average number of inelastic interactions per Bunch Crossing (BX)

$$n_{coll/bx} = L \times \sigma_{in.pp} \times 2.5 \times 10^{-8} \sim 25. \qquad (1.2)$$

1.3   The Compact Muon Solenoid

The CMS experiment is designed to enable searches for a wide variety of new physics. To do this, it must be able to perform high precision measurements of electrons, photons, and muons even in high $n_{coll/bx}$ (henceforth refered to as pileup) conditions.

Figure 1.1 is a depiction of a transverse slice through the CMS barrel. Within CMS, a superconducting solenoid with an internal diameter of 6 m provides an axially and uniform 3.8 T magnetic field within its volume. Within this volume a silicon pixel Inner Tracker (IT) and a silicon microstrip Outer Tracker (OT) are located. These trackers are surrounded by electromagnetic and hadronic calorimeters. The angular coverage is extended by forward calorimeters. Gaseous ionisation detectors, used to detect muons, are embedded within the magnet's return yoke.

A highly detailed description of the CMS detector can be found in [8]. The coordinate system used by CMS, and throughout this thesis is pictured in Fig. 1.2. The Interaction Point (IP) is at approximately $x = y = 0$. An additional commonly used spacial coordinate is pseudorapidity, $\eta = -\ln[\tan(\theta/2)]$.

1.3.1 Tracker

The CMS tracker must measure the trajectories of charged particles in three dimensions. The transverse momentum (p_T) of the particles can be calculated from their radius of curvature in the magnetic field. The CMS tracker is the largest all-silicon tracking detector in the word, with an active surface area of around 200 m^2. Silicon is well suited for the LHC environment, as it has been shown to be relatively radiation hard, has a fast charge collection time (which enables resolution of individual bunch crossings), and is structurally strong enough to allow the use of thin sensors, which reduce the energy losses and track deflections associated with detector interactions. An IT of silicon pixels is used up to a radius of 4.3 cm, where the flux of incident particles is much larger. A silicon microstrip OT is then used, up to a radius of 110 cm. Both of these trackers use a split barrel-endcap design, with a total of 13 barrel layers, and 14 endcap disks either side. The tracker extends to $|\eta| < 2.5$.

1.3.2 Electromagnetic Calorimeter

The CMS Electromagnetic Calorimeter (ECAL) was designed to measure the energy and position of electrons and photons produced either in the primary interaction, or in QCD jets. Fine granularity, radiation tolerant lead tungstate (PbWO$_4$) tapered crystals are used to induce electromagnetic cascades from traversing photons and electrons, resulting in scintillation light that undergoes total internal reflection towards the back of the crystal. Eighty percent of the light is collected within 25 ns. Photodiodes are required to amplify the relatively small signal (4.5 photoelectrons per MeV). The ECAL extends to $|\eta| < 3.0$.

The Electromagnetic Preshower (ES), a sampling calorimeter comprised of lead and silicon layers, is located in the forward region. Signals that look like high energy photons from the primary interaction can be generated when a π^0 decays into two

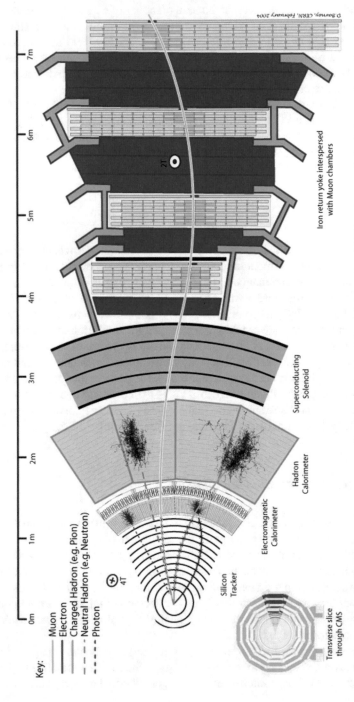

Fig. 1.1 Overview of the CMS detector, as a transverse slice through the barrel. It can be seen that CMS is constructed of sub-detector layers that are designed to measure particles produced in the LHC collisions

Fig. 1.2 Diagram of the coordinate system used by CMS, and in this thesis, illustrating the corresponding azimuthal (ϕ) and polar θ angles for a particle with momentum p, produced at the origin of CMS. The beam direction is parallel to the z axis, and collisions occur at approximately $x = y = 0$.

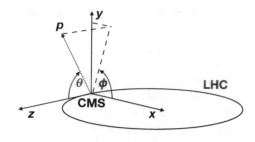

photons with a small separation angle. The ES has a much finer granularity than the ECAL (2 mm strips compared to 3 cm crystals), and is therefore able to resolve these photons individually.

1.3.3 Hadronic Calorimeter

The Hadronic Calorimeter (HCAL) encloses the ECAL, and is used to reconstruct the energy and position of QCD jets, particularly neutral hadrons. The HCAL extends to $|\eta| < 5.2$. HCAL Barrel Calorimeter (HB), and HCAL Endcap Calorimeter (HE) sampling detectors consist of layers of brass plates (which induce hadronic showers), and plastic scintillation tiles. The scintillation signal is collected and transferred to on-detector amplifiers. As a consequence of the higher particle flux in the forward regions, the HCAL Forward Calorimeter (HF) has been designed to use a more radiation-hard technology. In this subdetector steel attenuators containing quartz fibres are used to measure the Cherenkov radiation of a particle.

1.3.4 Muon Detectors

CMS is required to detect and identify muons with high efficiency, given their presence in the final states of many interesting decay channels (where the object of particular interest is the heavy state that has decayed). In CMS, three different gas chamber technologies are used: Drift Tubes (DT) in the barrel region; Cathode Strip Chambers (CSC) in the endcap regions; and Resistive Plate Chambers (RPC) for all $|\eta| < 2.1$. DTs are able to be used in the barrel region as the neutron-induced background is small, the muon rate is relatively low, and the magnetic field is uniform. In the endcap, however, where the muon and background rates are much higher, and the magnetic field is non-uniform, CSC are more suitable. To provide additional coverage in the scenario that background rates significantly rise as the LHC luminosity increases, an independent, highly segmented dedicated triggering system with a fast response time, consisting of RPC is also in use. In comparison with the other muon detectors, they deliver much better time resolution, but poorer position resolution.

1.3.5 Trigger and Data Acquisition

Due to constraints on data bandwidth and storage capacity, a trigger system is required to select interesting events for storage and analysis. A two-tier trigger system is used in CMS. The Level-1 (L1) trigger, is implemented in custom electronics, and is capable of reducing the event rate from 40 MHz to 100 kHz within the latency budget of around 4 μs (approximately 2 μs of this is required to transmit the data off-detector, and propagate back the L1 accept signal, including Serialisation/De-serialisation (SERDES) and other delays) [17–21]. The High Level Trigger (HLT) further reduces the event rate to about 1 kHz, using reconstruction and analysis software running on a farm of commercial computers. The HLT has access to the complete detector information; however, due to latency considerations (∼4 ms latency budget), simplified reconstruction algorithms are used when compared to full offline reconstruction. A sequential series of reconstruction and subsequent filter stages are executed, which terminate upon the failure to pass any given filter. This allows for fast pre-filters based on calorimeter information to reduce the number of events that must undergo more complex processing stages such as track reconstruction. A globally distributed data storage and processing infrastructure, known as the Worldwide LHC Computing Grid (WLCG) [22] is used to store and perform offline reconstruction on events that pass the HLT selection.

The current L1 trigger can be divided into two main subsystems, the L1 calorimeter trigger, and the L1 muon trigger. Each of these subsystems receives data from different subdetectors, and their outputs are combined in the Micro Global Trigger (MicroGT), which makes a final triggering decision. The outputs of the L1 muon trigger and L1 calorimeter trigger correspond to relevant physics objects: electrons and photons, tau lepton decays containing hadrons, jets, energy sums, and muons. Although the majority of trigger primitive processing and selection is done in the muon and calorimeter trigger layers, the MicroGT makes a decision based on the momentum, position, isolation, and quality of these objects using a set of around 300 algorithms, which are regularly updated to reflect changes in the LHC running conditions. The L1 trigger has been recently upgraded [23], to maintain its efficiency for the selection of interesting processes in the context of the increased pileup at LHC Run 2. This upgrade consisted of a complete replacement and commissioning of the electronics, simulation software, monitoring and configuration systems, databases, and the timing and data acquisition interface. It was installed and commissioned in 2015, and was successfully operated for the duration of 2016 and 2017 [24]. The ability to implement sophisticated triggering algorithms within the ∼4 μs latency budget was afforded by the use of a large Field Programmable Gate Array (FPGA) mounted on various custom data processing boards that conform to the Micro Telecommunications Computing Architecture (MicroTCA) standard (and Advanced Mezzanine Card (AMC) form-factor). These boards are equipped with up to 1 Tb/s worth of serial optical bandwidth each. One variety of these boards, the Master Processor Virtex-7 (MP7), will be described in Sect. 3.3.2.1.

The CMS Data Acquisition (DAQ) [25] must read out and assemble events accepted by the L1 trigger. A middleware framework known as XDAQ [26] (built in C++) allows the use of distributed data acquisition and high level triggering systems. XDAQ provides software for data transport, configuration, monitoring, and error reporting. The CMS Run Control and Monitoring System (RCMS) framework (written in Java) consists of a hierarchy of function managers which control XDAQ applications, and allows control of the detector via web applications.

The HLT, in addition to the offline reconstruction, analysis and simulation use the CMS Software (CMSSW). CMSSW is a framework based around an event data model, where each event is a C++ object container for all raw (or simulated) and reconstructed data related to a particular CMS bunch crossing or event. Individual processing modules may be run in isolation.

1.4 Field Programmable Gate Arrays

An FPGA is a semiconductor device, based on a matrix of configurable logic blocks, each connected via programmable interconnects. In contrast to an Application Specific Integrated Circuit (ASIC), which is custom manufactured to fulfil a specific task, FPGAs can be reprogrammed after manufacturing at the discretion of the user. Outside of High Energy Physics (HEP), FPGAs are used widely in many fields. Some examples include aerospace and defence, ASIC prototyping, high performance computing and data storage centers, wired and wireless network packet processing. Modern FPGAs contain Digital Signal Processors (DSP), Random Access Memory (RAM), and multi-gigabit transceivers. As they have not been designed for a specific use, in general FPGAs must be run slower, and with lower energy efficiency, than custom ASICs designed for a specific purpose.

In HEP, FPGAs have been proven to be extremely useful for fast data processing when workloads can be heavily pipelined or parallelised, in particular for trigger and DAQ operations. Utilisation of commercial FPGA technology allows increased flexibility, and reduced risk (and cost) when compared to developing an ASIC to do the same task. As a consequence of the required radiation hardness and low power dissipation, custom ASICs are still envisaged to be used for all aspects of the detector front-end; however, in the back-end processing, FPGAs are now the standard across HEP.

The behaviour of the FPGA is defined by the user in a Hardware Description Language (HDL) such as VHDL or Verilog (or occasionally with a schematic design). This code is often called FPGA *firmware*. An electronic design automation tool, usually provided by the FPGA vendors, is then used to generate a netlist that can be mapped onto the chip through a series of place, route, timing analysis and verification steps. The output of this process is a binary file that can be used to reconfigure the FPGA to the desired circuit. FPGA vendors and third parties often provide a library of verified and tested predefined circuits known as IP cores, which can be used for common tasks.

Several large corporations specialise in the design of FPGA chips. The majority of FPGAs used in HEP are designed by either Xilinx, Inc., or Altera Corporation (now a subsidy of Intel Corporation). This thesis presents work done with Xilinx FPGAs [27, 28], which are the standard adopted by the CMS tracker for future back-end data processing applications. The use of Altera devices [29, 30] has been considered, however they were ultimately rejected as a result of fewer high speed transceivers when compared to equivalently priced Xilinx counterparts. In addition, effort would be required to port existing Xilinx compatible firmware to Altera devices, as IP cores are not usually cross-compatible.

FPGAs may be categorised by their available logic resources. Firmware designs presented in this thesis will be given alongside their resource usage in terms of several important quantities. These quantities are as follows:

- DSPs are specialised microprocessing blocks that are designed to multiply and accumulate fixed-point bitwise data in parallel.
- Look Up Tables (LUT) are small memory units that can be used to implement an arbitrarily defined boolean function (such as a combination of logical AND, OR, XOR or NOT). In Xilinx devices a LUT has four boolean value inputs, the values of which can determine an output boolean value.
- Flip Flop (FF)s, also known as registers, are used to maintain the state of the FPGA between clocks, and keep the data moving in a synchronous way based on the edge of a digital clock.
- Block Random Access Memory (BRAM) is a dedicated two-port memory module containing 36 Kb of RAM each.

Further detail on the resources and functionality available within the currently available Xilinx Virtex 7, Xilinx Ultrascale, and Xilinx Ultrascale+ family of devices can be found in their respective Xilinx product data sheets [27, 28].

References

1. Martin BR, Shaw G (2008) Particle physics, 3rd ed. Manchester physics series. Wiley, ISBN-10:0470032944
2. Griffiths DJ (2008) Introduction to elementary particles, 2nd ed. Physics textbook, Wiley. ISBN 10:3527406018
3. Weinberg S (1967) A model of Leptons. Phys Rev Lett 19:1264–1266. https://doi.org/10.1103/PhysRevLett.19.1264
4. Evans L, Bryant P et al (2008) LHC Machine. JINST 3:S08001. https://doi.org/10.1088/1748-0221/3/08/S08001
5. Apollinari G et al (2015) High-Luminosity Large Hadron Collider (HL-LHC): preliminary design report, Dec 2015, CERN, Geneva. https://doi.org/10.5170/CERN-2015-005
6. ATLAS Collaboration (2008) The ATLAS experiment at the CERN Large Hadron Collider. JINST 3:S08003. https://doi.org/10.1088/1748-0221/3/08/S08003
7. ATLAS Collaboration (2012) Observation of a new particle in the search for the Standard Model Higgs boson with the ATLAS detector at the LHC. Phys Lett B 716:1–29, CERN-PH-EP-2012-218. https://doi.org/10.1016/j.physletb.2012.08.020

8. CMS Collaboration (2008) The CMS experiment at the CERN LHC. JINST 3:S08004. https://doi.org/10.1088/1748-0221/3/08/S08004

9. CMS Collaboration (2012) Observation of a new boson at a mass of 125 GeV with the CMS experiment at the LHC. Phys Lett B 716, CERN-PH-EP-2012-220. https://doi.org/10.1016/j.physletb.2012.08.021

10. Collaboration Super-Kamiokande (1998) Evidence for oscillation of atmospheric neutrinos. Phys Rev Lett 81(8):1562–1567. https://doi.org/10.1103/PhysRevLett.81.1562

11. Oort JH (1932) The force exerted by the stellar system in the direction perpendicular to the galactic plane and some related problems. Bull Astron Inst Netherlands 6:249–287

12. Planck Collaboration (2016) Planck 2015 results. XIII. Cosmological parameters, A&A 594 A13 page 1502.01589. https://doi.org/10.1051/0004-6361/201525830

13. Pape L, Treille D (2006) Supersymmetry facing experiment: much ado (already) about nothing (yet), (2006) Institute of Physics Publishing. Rep Prog Phys 69:2843–3067. https://doi.org/10.1088/0034-4885/69/11/R01

14. Jowet JM, Schaumann M, Alemany R et al (2016) The 2015 heavy-ion run of the LHC. In: Proceedings of IPAC 2016. Busan, Korea. https://doi.org/10.18429/JACoW-IPAC2016-TUPMW027

15. LHCb Collaboration (2008) The LHCb experiment at the CERN LHC. JINST 3:S08005. https://doi.org/10.1088/1748-0221/3/08/S08005

16. ALICE Collaboration (2008) The ALICE experiment at the CERN LHC. JINST 3:S08002. https://doi.org/10.1088/1748-0221/3/08/S08002

17. Kluge A, Smith WH (1996) CMS Technical Note CMS Level 1 Trigger Latency, Mar 1996, CMS TN/96-033

18. CMS Collaboration (2017) The CMS trigger system. JINST 12:P01020. https://doi.org/10.1088/1748-0221/12/01/P01020

19. CMS Collaboration (2010) Performance of the CMS Level-1 trigger during commissioning with cosmic ray Muons and LHC beams. JINST 5:T03002. https://doi.org/10.1088/1748-0221/5/03/T03002

20. Tapper A, Acosta D et al (2013) CMS technical design report for the Level-1 trigger upgrade, Jun 2013, technical report CERN-LHCC-2013-011

21. CMS Collaboration (2000) CMS TriDAS project: technical design report, volume 1: The trigger systems, Dec 2000, technical report CERN-LHCC-2000-038

22. CMS Collaboration (2005) CMS: The computing project, Jun 2005, technical report CERN-LHCC-2005-023

23. CMS Collaboration (2013) CMS technical design report for the Level-1 trigger upgrade, Jun 2013, technical report CERN-LHCC-2013-011

24. Zabi A et al (2017) The CMS level-1 trigger system for LHC Run II. JINST 12:C01065. https://doi.org/10.1088/1748-0221/12/01/C01065

25. CMS Collaboration (2002) The TriDAS project, technical design report, volume 2: data acquisition and high-level trigger, Dec 2002, technical report CERN/LHCC 02-26

26. Berti L et al (2003) Using XDAQ in application scenarios of the CMS experiment, Mar 2003, Computing in High-Energy and Nuclear Physics. Journal ref: ECONFC0303241:MOGT008

27. Xilinx Inc (2017) 7 series FPGAs data sheet: overview, Aug 2017, product specification, DS180 (v2.5)

28. Xilinx Inc (2017) UltraScale architecture and product data sheet: Overview, Feb 2017, product specification, DS890 (v2.11)

29. Intel Corporation (2018) Intel Arria 10 device overview, Jan 2018, A10-OVERVIEW

30. Intel Corporation (2017) Stratix 10 GX/SX device overview, Oct 2017, S10-OVERVIEW

Chapter 2
The CMS Phase II Upgrade

2.1 The High-Luminosity LHC

In order to fully exploit the scientific potential of the LHC, it is planned to operate the machine at a higher average instantaneous luminosity, $L = 5$–$7.5 \times 10^{34}/\text{cm}^2/\text{s}$ from 2026 onwards, following a 30 month-long shut down. This luminosity is equivalent to 140–200 proton-proton collisions per 40 MHz bunch crossing. The total integrated luminosity target of 3000–4000 fb^{-1} by 2038 should allow more precise measurements of the Higgs boson couplings, and an improved discovery reach for new particles with multi-TeV masses, and/or low cross-sections [1].

2.2 Motivation for an Upgraded CMS Tracker

Long Shutdown 3 (LS3) is a 30 month shut down of the LHC, scheduled to last from 2024 to 2026, in which the machine upgrades that allow high-luminosity running will be installed. At this same time, major installation work is planned for the LHC detectors, both for required maintenance and upgrade. Alongside many other upgrades, the CMS detector will be installing a new silicon tracker (CMS Phase II Outer Tracker) [2, 3], a new pixel detector (CMS Phase II Inner Tracker) [2, 3], and a High Granularity Calorimeter (HGCal) [4] which will replace the current endcap ECAL.

During the shutdown commencing in 2024, the CMS tracker must be completely replaced, primarily due to the accumulation of radiation damage to the silicon sensors by this time. Maintaining the current track reconstruction performance under the increased pileup conditions of the HL-LHC will be a significant challenge when designing the new tracker. To keep the misidentification (fake) rate low, the new tracker will need a finer sensor channel granularity. It must also be significantly more radiation hard [5]. The offline tracking efficiency and fake rate for $t\bar{t}$ event tracks with a $p_T > 0.9$ GeV are shown in Fig. 2.1. The blue markers represent the expected performance if the tracker was not replaced. By LS3 it will have received radiation damage from the 300 fb^{-1} delivered by the LHC by that time. In addition, the higher pileup value of 140 significantly increases the misconstructed track rate.

© Springer Nature Switzerland AG 2019
T. O. James, *A Hardware Track-Trigger for CMS*, Springer Theses,
https://doi.org/10.1007/978-3-030-31934-2_2

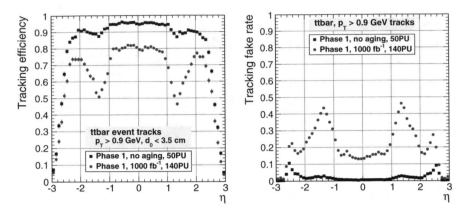

Fig. 2.1 Tracking efficiency (left) and fake rate (right) for the current CMS tracker at 50 pileup without ageing (black), and at 140 pileup with ageing (blue) [2]

It is clear from Fig. 2.1 that significant performance reduction in comparison with the present operation would occur if it was not replaced in time.

A major new feature that has shaped the design of the new tracker is the ability to read out some (limited) tracking information at 40 MHz to the L1 trigger. Data from the Outer Tracker could be used to keep the L1 acceptance rate below the expected 750 kHz maximum, without loss in sensitivity to interesting physics. Simulations suggest that the provision and use of tracking information at L1 in the form of fully reconstructed tracks with $p_T > 3$ GeV is a necessity if trigger performance is to be maintained or improved upon relative to the low luminosity conditions. It is estimated that under a high pileup scenario (200 pileup), the L1 rate would not exceed the 750 kHz maximum if tracks were used to enhance the discrimination power of the trigger for a given set of p_T or energy thresholds. Conversely, without the use of tracks, but under the same conditions, the L1 rate would be expected to exceed 4 MHz [3, 6].

2.3 The Phase II Outer Tracker Design and Geometry

Figure 2.2 shows two proposed Phase II Outer Tracker geometries [2]. The Outer Tracker provides radial coverage in the region $21 < r < 112$ cm. Six cylindrical barrel layers cover the region $|z| < 120$ cm and five endcap disks on each side provide coverage up to $|z| < 270$ cm. This configuration was chosen to ensure that tracks originating within the expected luminous region (where the beams intersect) will pass through a minimum of six tracking layers up to about $|\eta| = 2.4$. This is required to ensure robust track finding performance at L1, even in the event of one such layer becoming inefficient.

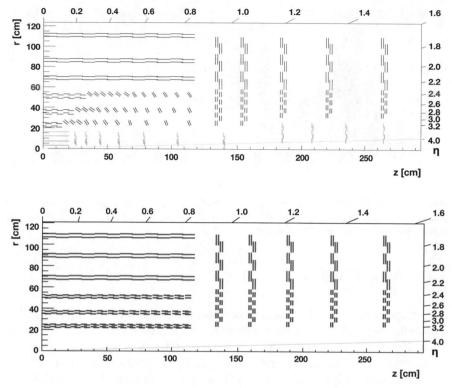

Fig. 2.2 A *tilted barrel* (upper) and *flat barrel* (lower) design of one quadrant in the $r - z$ plane of a proposed upgraded Outer Tracker layout, showing the 2S (red) and PS (navy) module placement. The tilted barrel map shows the (fully pixel) Inner Tracker in yellow and light blue

Performance simulations (including simulated physics event samples used throughout this thesis) rely on Molière's formula [7, 8] for the RMS width of the central 98% of the projected multiple Coulomb scattering angle. For a particle with velocity β, momentum p [GeV/c], and charge q:

$$\theta_{\text{RMS}}^2 = \left(\frac{13.6\, q\, \text{MeV}}{\beta c p} \right)^2 \frac{x}{X_0} \left[1 + 0.038 \ln (x/X_0) \right]^2 , \qquad (2.1)$$

where x is the thickness of the material, with radiation length X_0, traversed. Total multiple scattering errors are estimated by adding the contributions from each layer in quadrature. Fewer tracking layers, thinner sensors (about $300\,\mu$m compared to $500\,\mu$m in the OT), and improved service routing allow the Phase II tracker to contain significantly less material than the present tracker, as shown in Fig. 2.3. The method and parameters for these results, and the Monte Carlo simulation of LHC events in the Phase II tracker are described in more detail in [2].

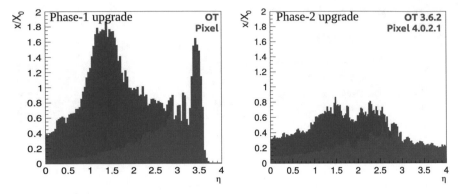

Fig. 2.3 Material budget of the current (Phase I) tracker (left) and the proposed Phase II tracker with tilted barrel geometry (right) [2]. Material in the Inner Tracker is shown in red, and material in the outer tracker is in blue. The histograms are stacked

2.4 The p_T-Modules

A new type of tracker module has been developed to allow real-time track-triggering despite the bandwidth limitations that prevent every tracker hit to be read off-detector at 40 MHz [9–11]. The design of the new modules, (known as p_T-modules), exploits the knowledge that tracks with high p_T are usually most relevant to physics reconstruction. Two closely spaced silicon sensors, bonded to a single read-out ASIC, are used to record pairs of clusters within correlation windows compatible with a high p_T (low bend) charged particle. This is depicted in Fig. 2.4.

The correlation windows and sensor separation vary depending on radius and whether the module is in the barrel or endcap. In addition, two main types of modules are being produced, each optimised for use at different radii. The parameters of these modules are shown in Table 2.1. Two-Strip (2S) modules are designed to operate at $r > 60$ cm, and consist of two sensors (upper and lower), each containing two sets (left and right) of 1016 silicon strips of size 5 cm × 90 μm. Each sensor consists of two isolated half-sensors, allowing half-module granularity in the direction transverse to the strip length. The Pixel-Strip (PS) modules are designed to operate at the higher occupancy region $r < 60$ cm. They consist of one sensor of silicon macro-pixels to provide finer granularity in the direction approximately transverse to the particle trajectory (the z direction in the barrel, and the r direction in the endcaps), and one sensor of silicon strips. These modules are designed to operate at −20 °C. In total, there are expected to be 7680 2S and 5616 PS modules in operation. It should be noted that stub data will not be sent from modules located at $|\eta| > 2.4$, as tracks at these pseudorapidity would leave hits in too few layers for L1 track finding.

Each p_T module will be served by one upstream and one downstream transmitting optical fibre, which directly interfaces to the first layer of back-end electronics. Depending on the module radius, these links will be capable of transferring data off-detector at either 5.12 or 10.24 Gb/s, providing an effective bandwidth of be-

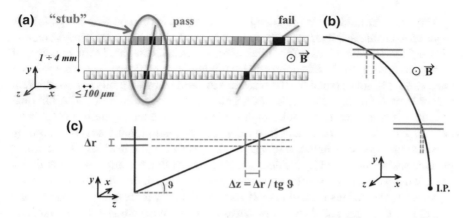

Fig. 2.4 Cluster matching in p_T-modules to form stubs. **a** Correlating closely spaced clusters between two sensor layers, separated by a few mm, allows discrimination of p_T based on the particle bend in the CMS magnetic field, assuming that the particle originated at the beam-line. **b** For a particle originating at the IP, the same p_T corresponds to a larger distance between signals at larger radii, for a given sensor spacing. **c** A larger spacing is needed in the endcap disks to achieve the same discrimination power. The required spacing in z is given by the spacing in r divided by $\tan\theta$, where θ is the polar angle. Only stubs compatible with tracks with $p_T > 2$–3 GeV are transferred off-detector[2]

Table 2.1 Parameters of two-strip and pixel-strip p_T-modules. The silicon strip sensors are split into two isolated halves, whose front end chips cannot communicate

	2S module	PS module
Active area	$2 \times 90\,cm^2$	$2 \times 45\,cm^2$
Sensor one	2×1016 strips	2×960 strips
Sensor two	2×1016 strips	32×960 macro-pixels
Strip size	$5\,cm \times 90\,\mu m$	$2.4\,cm \times 100\,\mu m$
Macro-pixel size	N/A	$1.5\,mm \times 100\,\mu m$
Front-end power [W]	5	8
Sensor power at $-20\,°C$ [W]	1.0	1.4

tween 3.84 and 8.96 Gb/s allowing for error correction and protocol overheads[12]. Approximately 75% of this bandwidth will be dedicated to readout of stub data from bunch crossings every 25 ns. The stub data format itself is dependent on the p_T-module type, but will contain an 11-bit address corresponding to the mean location of the hits in the seed cluster (to the nearest half-strip), and a 3-bit (PS) or 4-bit (2S) number, which corresponds to the distance in strips between the two clusters in the stub, or the local bend of the particle trajectory. For PS modules only, a 4-bit address describing the position of the stub along the sensor in the direction perpendicular to the strip granularity is additionally provided. The remaining approximately 25% of the module readout bandwidth will be dedicated to transmission of the full event data including all hit strips/pixels, triggered by a L1-accept signal[2].

The tilt of the modules in the three PS barrel layers (as can be seen in Fig. 2.2), is required to ensure that their normals point towards the interaction (luminous) region. This significantly improves stub-finding efficiency when compared to an entirely flat barrel geometry as tracks that approach the module at an angle may cross both sensor halves. As it is not possible for the front end to form stubs with hits in two different sensor halves, this particle would not generate a stub regardless of p_T. Additional benefits of using tilted PS modules in this region are a reduced system cost (as there are fewer modules in total), and fewer unnecessary, additional stubs, produced by overlapping modules within a single layer. A legacy design known as the *flat barrel* geometry, also shown in Fig. 2.2 does not have these tilted modules. As the majority of the work was done before the newer design was selected, the legacy design is used for some studies and results within this thesis, particularly for results that required the longer lead time associated with hardware and firmware development. Software and emulation studies are done using the tilted geometry scenario wherever possible. To facilitate track finding studies, stub-finding logic between the two module halves is enabled for flat barrel simulations, but this would not be possible in the final system.

2.4.1 Front-End Electronics

On-module ASIC chips are under development for use in the 2S and PS modules. The chips are designed for either 130, or 65 nm CMOS technology. The 65 nm feature size is more expensive, but offers more digital logic for the same footprint and power draw, and is therefore required for the PS module front-ends. Specialised radiation-hard chips are needed to perform a variety of on-module applications:

- The CMS Binary Chip (CBC), is the front-end chip of the 2S modules [13, 14]. Each module hosts sixteen CBC chips, which in turn each process data from 127 strips from each sensor (254 strips in total each). The CBC is designed to find clusters of hits with a programmable maximum width, and match clusters between the top and bottom sensors to produce stub data at a 40 MHz rate. In the case of an L1 acceptance, the CBC will provide non zero-suppressed hit data in binary form. The chip is designed with a peaking time of 20 ns, and a return to baseline within 50 ns. It dissipates about 130 mW.
- The Concentrator Integrated Circuit (CIC) [15] buffers, zero-suppresses and aggregates the data from all eight CBCs per half-module. It will deliver the trigger data in 8 BX synchronous blocks, allowing the limited bandwidth to be shared across time, such that local statistical fluctuations in hit rate can be smoothed. The same concentrator chip will be used to transfer data from the PS modules. It must therefore be a dual speed chip, that can be configured to match either 5.12 Gb/s or 10.24 Gb/s requirements.
- The Strip Sensor ASIC (SSA) is one of two 65 nm CMOS chips designed for the PS module front-end. The SSA processes the signals from the PS module strip

sensors, and sends zero-suppressed cluster data at 40 MHz to the corresponding macro-pixel front-end chip, the Macro Pixel ASIC (MPA).

- Sixteen MPA [16] chips are bump-bonded to each macro-pixel sensor. The resulting Macro-Pixel Sub-Assembly (MaPSA) contains approximately 30,000 bump-bonded macro-pixels. The MPA processes and zero-suppresses the hits from each connected macro-pixel. It then correlates the hits with those received from the corresponding SSA chip, to produce trigger stubs at 40 MHz. The MPA is designed for a peaking time of 24 ns, and a return to baseline within 50 ns.

Additionally, the CMS tracker will take advantage of common electronics developments at CERN. The following chips are examples of this:

- The Low-power Gigabit Transceiver (LpGBT) [12, 17] radiation hard serialiser will provide up to 10.24 Gb/s up links (with the option of 5.12 Gb/s in exchange for reduced power consumption), and 2.56 Gb/s down links, and will be qualified for a total fluence of 2×10^{15} n_{eq}/cm^2. The LpGBT must also be Single Event Upset (SEU) robust. In the tracker, the LpGBT will act as the I2C master of the modules, and will therefore control, monitor and configure the front-end ASICs.
- The Versatile Link Plus (VL+) [18] will be a radiation hard optical link driver that matches the up and down bandwidth of the LpGBT.
- DC-DC converters will be mounted on the modules to deliver the input voltages required by the front-end electronics. This allows ohmic losses to be minimized in the supply cables.

In contrast to the original CMS tracker, the front-end will deliver only binary data. This is required to reduce the data size in the high pileup environment, so that it can be read-out within bandwidth limitations. As a consequence, with the exception of the analogue front-ends, the electronics system will be fully digital. Binary data also negates the power requirements for detector Analog-to-Digital Converter (ADC)s, which are estimated to consume 1 mW per channel. The design has been shaped to account for a four-to-one ratio of trigger to DAQ data.

2.4.2 Sensor Type

In general, n-on-p type silicon sensors will be used. Measurements suggest that sensors with holes (in contrast to electrons) for signal generation suffer a greater degradation in charge collection following radiation damage. Radiation surface damage increases inter-strip capacitance, and therefore noise, for p-on-n type sensors. In contrast, surface damage causes signal sharing between strips for n-on-p type sensors, but this is mitigated by a p-doped structure surrounding the strips, which isolates them. Simulations and radiation tests have shown that the chosen sensor types will be able to maintain performance after radiation fluence, meeting (or exceeding) what they will be subject to at the HL-LHC [3, 19].

2.5   Module Prototyping and Beam Tests

Both 2S and PS module prototypes have been developed to evaluate performance and robustness. These prototypes have been tested in fixed target beam facilities. Beam facilities allow the verification of the stub-finding performance of the modules, both before and after significant radiation exposure.

One such prototype module is the 2S mini-module, shown in Fig. 2.5. This module consists of two stacked 5 cm (254 strip) long strip sensors on an aluminium frame, wire-bonded to two prototype CBCs. Three full-sized 2S module prototypes have also been assembled at CERN. They contain two full size 2S sensors (10 cm, 1016 strips), connected to sixteen prototype CBCs.

A prototype MAPSA (the MAPSA-light), and a PS micro-module consisting of two stacked MAPSA-light assemblies with approximately 3.3 mm sensor separation have also been assembled [2]. The MAPSA-light [20] is a bump bonded assembly of up to six small prototype MPAs, each with 48 readout channels, to a small macro-pixel sensor of size 1.2 cm × 0.78 cm.

2.5.1   Test Beam Apparatus

Test beam experiments for the 2S prototype modules utilised the (EU-funded) AIDA telescope [21], a tracking detector comprising six planes of silicon sensors for accurate track reconstruction. In addition, a fast-timing reference plane [22] consisting of a pixel sensor bump bonded to the ATLAS pixel readout chip (the FE-I4) is used for accurate timing resolution, and a pair of crossed scintillators located at either end of the telescope are used for trigger generation. The synchronization of the data streams from the 2S module, the sensor planes, and the fast timing plane is performed

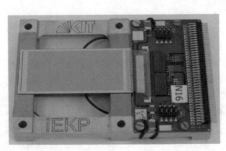

Fig. 2.5 (Right) A full sized 2S module prototype (1.8 mm sensor spacing variant), equipped with eight CBC prototype chips, and two 10 cm long silicon strip sensors [2]. (Left) A 2S mini-module, assembled from a small prototype hybrid containing two prototype CBC readout chips, and two 5 cm long strip sensors mounted in an aluminium frame [2]

by an FPGA-based Trigger Logic Unit (TLU) [23]. Dedicated NIM logic is used to generate the trigger signal from the output signals of the two pairs of crossing scintillators at either end of the telescope. This signal is provided as input to the TLU, which distributes it to the module under test and the telescope sensor planes. A simple handshake protocol is used to maintain synchronization between the different detector systems during data taking.

As the telescope outputs a 112 µs rolling-shutter frame (the charge is measured over the full shutter time, and is read out by scanning across strips, meaning that not every strip is measured simultaneously), the fast timing plane is required to correlate the multiple tracks in the frame to the individual triggers. The fast timing plane runs at 40 MHz, allowing the required time resolution of 25 ns to be achieved. The fast timing plane also allows a sensitive region of interest to be configured, increasing the fraction of DAQ events that correspond to tracks crossing the device under test.

The DAQ system for the CBC module tests use the CERN Gigabit Link Interface Board (GLIB) MicroTCA board [24]. As in the final system, communication between the back-end and the front-end is achieved via an optical fibre connection for long distance transmission at 4.8 Gb/s using the GigaBit Transceiver (GBT) protocol. The data received from the CBCs are processed and formatted by the firmware and then sent to a XDAQ application that formats events in a CMS-compatible format, and stores the data for later processing. This is done using a standard CMS online software chain that provides a file format compatible with the CMS reconstruction software, CMSSW. The DAQ architecture for future beam test experiments with the latest prototype CBC, MPA and SSA chips will be based on the newer FC7 AMC card [25], which features one Kintex 7 XC7K480T FPGA [26], providing additional logic resources.

2.5.2   Test Beam Results

Beam test experiments can be used to validate the performance of the prototype modules. One important metric is the stub reconstruction efficiency, and its dependence on beam incidence angle, which can be used to emulate the bending of tracks in the magnetic field of CMS. The stub reconstruction efficiency is defined as the ratio between the number of events with at least one stub that matches a reconstructed track, and the number of events with one track matched to the hits in the trigger planes. The track-stub matching requires that the track and stub positions must match to within 4σ, accounting for the spatial resolution of the devices.

The radius of curvature R (in m) of a charged particle with transverse momentum p_T (in eV), and charge q (in electron units), bent in the transverse plane by a homogeneous magnetic field of strength B (in Tesla) is given by,

$$R = \frac{p_T}{qBc}. \tag{2.2}$$

Fig. 2.6 Illustration of
relationship between
incident beam angle on a
tracker module, and radius of
curvature of traversing
particle from the interaction
point (I.P.). For a module at a
given radius, r, the angle of
incidence with respect to the
perpendicular of the module,
α, decreases with an
increased radius of
curvature, R (which
corresponds to an increase in
p_T). As the angles marked α
are equivalent, one can
therefore construct Eq. 2.3

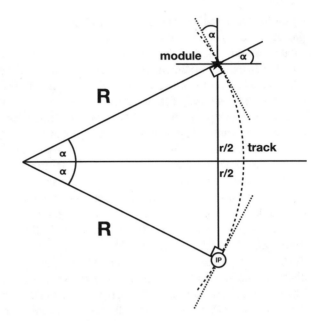

Using Fig. 2.6, one can write down the relation

$$\sin \alpha = r/2R. \tag{2.3}$$

By combining this result with Eq. 2.2, one can derive the relationship between
beam incident angle, α, and the p_T of a traversing particle at radius r, within the
CMS magnetic field

$$p_T[\text{eV}] = \left(\frac{qBc}{2}\right)\frac{r}{\sin \alpha}, \quad p_T[\text{GeV}] = \frac{0.57 \times r}{\sin \alpha} \tag{2.4}$$

The stub turn on curve, the increase in stub reconstruction efficiency from 0 to
close to 1 at around $p_T = 2\,\text{GeV}$ is shown in Fig. 2.7, for both irradiated and non-
irradiated 2S mini-modules. The irradiated mini-module was subject to a fluence of
$6 \times 10^{14}\,\text{n}_{\text{eq}}/\text{cm}^2$, approximately twice that expected during lifetime operation at
the HL-LHC. These measurements were taken at the CERN H6B beam line, using
120 GeV pions. One can see that for the non-irradiated module, the turn-on threshold,
which is defined as the p_T for which stub efficiency reaches 50%, was found to
be 2.16 GeV. The sharp turn on, and a plateau efficiency of 99% is a successful
demonstration of the module's target functionality to efficiently select stubs above
the chosen threshold. In the case of the irradiated mini-module, a plateau efficiency of
above 95% is observed within the beam core (assuming the module is perpendicular
to the incoming particle in the r-z plane) [2].

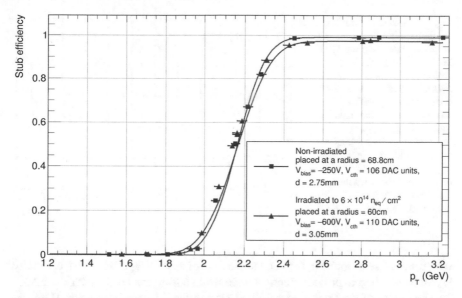

Fig. 2.7 The stub turn on curve. Stub reconstruction efficiency for an irradiated (blue), and non-irradiated (red) 2S mini-module, as a function of p_T [2]. Irradiation fluence was $6 \times 10^{14}\,\mathrm{n_{eq}/cm^2}$. The threshold setting V_{cth}, sensor spacing d, and bias voltage V_{bias} are given in the legend for each module. The V_{cth} values used correspond to about 4900 and 3500 electrons for the unirradiated, and irradiated module respectively As the modules have different sensor spacing, different radii (68.8 and 60 cm) were used to calculate a p_T from the angle of incidence, using Eq. 2.4

The average stub reconstruction efficiency was measured with a full-size prototype 2S module, per CBC chip. Across all chips, (with 2 out of 1000 strip pairs masked), the average efficiency is measured to be 97.3%. This is approximately 2% lower than that observed in the 2S mini-module test, which is a result of a contamination of events where the module and telescope were not synchronised.

As can be seen in Fig. 2.8, the expected dependence on p_T and cluster matching windows Δx (in macro-pixel units) is observed in MAPSA-light beam tests at FNAL. For a value of $\Delta x \leq 4$, a resolution of 6.5% is measured at 2.2 GeV.

2.6   Back-End Electronics

The proposed data path for the upgraded tracker is depicted in Fig. 2.9. The first layer of off-detector readout electronics will be the Data, Trigger, and Control (DTC) boards. These custom-developed boards will comply with the ATCA form-factor [27], and will contain commercial FPGAs, and opto-electronics transceivers. This board will be designed to pre-process the stub data before transmission to a downstream track finder layer. It must also extract and package the full event data,

Fig. 2.8 Stub reconstruction efficiency of a PS micro-module, the prototype MAPSA, at a 120 GeV proton test beam, at FNAL [2]. The denominator is determined by tracks reconstructed offline. This fraction is given as a function of p_T (at a reference radius of 51.7 cm), for values of Δx, the number of macro-pixels between the clusters in the top and bottom module sensor

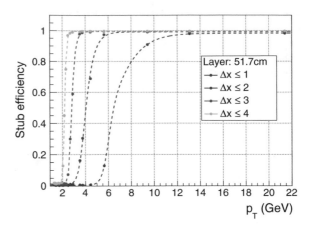

which will be sent from the front-end in the case of an L1 trigger acceptance. Finally, it will be required to provide timing and control paths to the on-detector modules.

The output of the track finder will be sent to the L1 correlator layer. Here, the tracks will be combined with the trigger primitives from other sub-detectors, and a global trigger decision will be formed. In order to make the trigger decision, and relay that information back to the front-end (which has a finite event buffer that must be triggered within 12.5 μs) in time, it is anticipated that the L1 correlator will need the tracks within 5 μs of the collision. This number is estimated by budgeting 0.5 μs for each of the following operations, based on experience with similar algorithms of binning, sorting and routing fixed point quantities of a certain size, as implemented in the current trigger: primary vertex finding; matching tracks with calorimeter and the muon objects; calculating a new L1 object with the combined information; calculating the isolation of the muon and calorimeter objects [3]. A further 1 μs is budgeted for global processing such as global sums, kinematic calculations and trigger logic. The propagation of the trigger decision back to the front-end will require another 1 μs (including SERDES and other delays). A safety margin of 30% is then applied, the amount by which the original CMS trigger exceeded its target latency. As one must also budget 1 μs for the stubs to arrive at the DTC, the stub pre-processing and track reconstruction and fitting must be accomplished within 4 μs.

Each track finder and DTC crate must also be equipped with a common card, the DAQ and Timing Hub (DTH). The DTH will be an ATCA card, and will provide timing and control distribution, as well as the DAQ path for the leaf cards in the crate. It will convert the data to a commercial network protocol such as 100 Gb/s (4 × 25 Gb/s lanes) Ethernet.

The following chapters will present a detailed discussion of the functionality and scope of the track finding layer, which consists of a number of Track Finding Processor (TFP) boards. This thesis will present what has been achieved to date in demonstrating the feasibility and performance of a such an object, with currently available technology.

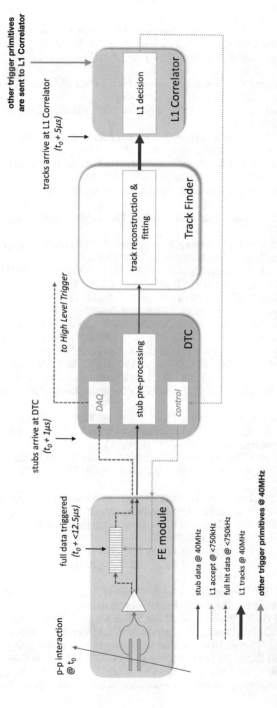

Fig. 2.9 Illustration of dataflow and latency requirements from p_T-modules through to the off-detector electronics dedicated to forming the L1 trigger decision [28]

References

1. Apollinari G et al (2015) High-Luminosity Large Hadron Collider (HL-LHC): preliminary design report, Dec 2015, CERN, Geneva. https://doi.org/10.5170/CERN-2015-005
2. CMS Collaboration (2017) The Phase-2 upgrade of the CMS tracker, Jul 2017, technical report CERN-LHCC-2017-009
3. CMS Collaboration (2015) Technical proposal for the Phase-II upgrade of the CMS detector, Jun 2015, technical report CERN-LHCC-2015-010
4. CMS Collaboration, The Phase-2 upgrade of the CMS endcap calorimeter, technical report CERN-LHCC-2017-023
5. Abbaneo D (2016) Performance requirements for the phase-2 tracker upgrades for ATLAS and CMS. EPJ Web Conf 127:00002. https://doi.org/10.1051/epjconf/201612700002
6. CMS Collaboration (2017) The Phase-2 upgrade of the CMS L1 trigger interim technical design report, Sep 2017, technical report CERN-LHCC-2017-013
7. Bethe HA (1952) Molière's theory of multiple scattering. Phys Rev 89(6). https://doi.org/10.1103/PhysRev.89.1256
8. Patrignani C et al (Particle Data Group) (2016) The review of particle physics. Chin Phys C 40:100001
9. Pesaresi M, Hall G (2010) Simulating the performance of a pT tracking trigger for CMS. JINST 5:C08003. https://doi.org/10.1088/1748-0221/5/08/C08003
10. Hall G, Raymond M, Rose A (2010) 2-D PT module concept for the SLHC CMS tracker. JINST 5:C07012. https://doi.org/10.1088/1748-0221/5/07/C07012
11. Pesaresi M (2010) Development of a new Silicon Tracker for CMS at Super-LHC, Jan 2010, Imperial College London PhD thesis, CERN-THESIS-2010-083
12. Moreria P et al (2009) The GBT project, Sep 2009, Topical Workshop on Electronics for Particle Physics, Paris, France, pp 342–346. https://doi.org/10.5170/CERN-2009-006.342
13. Raymond M, Braga D, Ferguson W et al (2012) The CMS binary chip for microstrip tracker readout at the SLHC. JINST 7:C01033. https://doi.org/10.1088/1748-0221/7/01/C01033
14. Braga D, Hall G, Jones L et al (2012) CBC2: a microstrip readout ASIC with coincidence logic for trigger primitives at HL-LHC. JINST 7:C10003. https://doi.org/10.1088/1748-0221/7/10/C10003
15. Caponetto L, Viret S, Zoccarato Y (2015) CIC1 technical specification, Dec 2015, Institut de Physique Nucléaire de Lyon (FR). https://espace.cern.ch/Tracker-Upgrade/Electronics/CIC/Shared%20Documents/Specifications/CIC_specs_v1.1.pdf
16. Ceresa D et al (2014) Macro Pixel ASIC (MPA): the readout ASIC for the pixel-strip (PS) module of the CMS outer tracker at HL-LHC. JINST 9:C11012. https://doi.org/10.1088/1748-0221/9/11/C11012
17. Moreira P (2017). LpGBT specification document, Jul 2017. https://espace.cern.ch/GBT-Project/LpGBT/Specifications/LpGbtxSpecifications.pdf
18. Soós C et al (2017) Versatile link PLUS transceiver development. JINST 12:C03068. https://doi.org/10.1088/1748-0221/12/03/C03068
19. König A, Bergauer T, Dragicevic M, Humann B, (2017) Field effect transistor test structures for p-stop strip isolation studies. JINST 12:C02067. https://doi.org/10.1088/1748-0221/12/02/C02067
20. Grossmann J (2017) PS-module prototypes with MPA-light readout chip for the CMS tracker phase 2 upgrade. JINST 12:C02049. https://doi.org/10.1088/1748-0221/12/02/C02049
21. Jansen H et al (2016) Performance of the EUDET-type beam telescopes, May 2016, EPJ Techniques and Instrumentation 3. https://doi.org/10.1140/epjti/s40485-016-0033-2
22. Obermann T et al (2014) Implementation of a configurable FE-I4 trigger plane for the AIDA telescope. JINST 9:C03035. https://doi.org/10.1088/1748-0221/9/03/C03035
23. Cussans D (2009). Description of the JRA1 Trigger Logic Unit (TLU), Sep 2009, v0.2c, EUDET-Memo-2009-4.   https://www.eudet.org/e26/e28/e42441/e57298/EUDET-MEMO-2009-04.pdf

24. Vichoudis P et al (2010) The Gigabit Link Interface Board (GLIB), a flexible system for the evaluation and use of GBT- based optical links. JINST 5:C11007. https://doi.org/10.1088/1748-0221/5/11/C11007

25. Pesaresi M et al (2015) The FC7 AMC for generic DAQ & control applications in CMS. JINST 10:C03036. https://doi.org/10.1088/1748-0221/10/03/C03036

26. Xilinx Inc (2017) 7 series FPGAs data sheet: overview, Aug 2017, product specification, DS180 (v2.5). https://www.xilinx.com/support/documentation/data_sheets/ds180_7Series_Overview.pdf

27. PICMG (2003) AdvancedTCA Short Form Specification, Jan 2003. https://indico.cern.ch/event/119030/attachments/61294/88092/PICMG_3_0_Shortform.pdf

28. Aggleton R et al (2017) An FPGA based track finder for the L1 trigger of the CMS experiment at the High Luminosity LHC. JINST 12:P12019. https://doi.org/10.1088/1748-0221/12/12/P12019

Chapter 3
The Track Finder Demonstrator

From LHC Run 4, CMS will require the use of fully reconstructed tracks with $p_T > 3\,\text{GeV}$ to maintain or improve trigger performance with respect to running at design LHC luminosity. Simulations suggest that the use of fully reconstructed tracks as an additional discriminatory handle could reduce the L1 accept rate at a 200 pileup scenario from 4 MHz to below 750 kHz. Trigger requirements will dictate whether the track finder must also reconstruct tracks down to as low as 2 GeV. This decision must be made on a trade-off between physics potential and the financial cost required to process and read out the increased number of stubs.

3.1 L1 Tracking Requirements

In order to ensure that the L1 tracks are useful to the L1 trigger, they must meet certain criteria of efficiency, purity, and resolution.

To facilitate single lepton triggers, the highest possible efficiency should be targeted for $p_T > 20\,\text{GeV}$ isolated electron and muon tracks. A high efficiency for lower p_T leptons is also required for dilepton triggers. Jet vertexing (the determination of a vertex position from which the jet originates) is required for the trigger to perform well in the high pileup conditions expected at HL-LHC. For this reason, the L1 track finder must find a substantial enough number of charged hadrons in each jet to determine the jet vertex. Higher p_T tracks in jets are particularly important for track-based missing transverse energy triggers. A low fake track rate at L1, especially at high p_T, is also important for these triggers as they would generate false signals. For use in the trigger it is important that tracking works well up to $\eta < 2.4$. The isolation performance degrades by about 5% when tracks are reconstructed only down to 3 rather than 2 GeV [1]. As the tracker is designed to measure particle p_T with high precision, the L1 p_T resolution will be better than is required for muon selection. Similarly, η and ϕ_0 resolution must be precise enough to match tracks with muon trigger and calorimeter information, but full tracker resolution may not be required.

© Springer Nature Switzerland AG 2019
T. O. James, *A Hardware Track-Trigger for CMS*, Springer Theses,
https://doi.org/10.1007/978-3-030-31934-2_3

For pileup mitigation it is important that the L1 track finder is able to achieve a resolution of a few mm in z_0.

3.2 Proposed Track Finder System Architecture

To be suitable for use in CMS, a track finding system must be scalable, configurable, and have redundancy. For this reason, a highly time-multiplexed design is desirable [2, 3].

3.2.1 A Time-Multiplexed Trigger

Time-multiplexing is the concept of buffering data from multiple sources, but belonging to a single event, such that it can be transmitted to a single processor node, over a longer period of time. Processing of sequential events can therefore be carried out on parallel processing nodes (each corresponding to a single event). An example of time-multiplexing, as used in the CMS trigger, is the L1 calorimeter trigger [4]. For a fixed time-multiplexing factor of P events, one would require P nodes, where each node processes a new event every $P \times \Delta P$ seconds, where ΔP is the time between subsequent events; 25 ns at the LHC. In general, a time-multiplexed design eliminates boundaries in hardware, which is especially advantageous for algorithms that require a large field of view, or a large number of interconnects between trigger regions. A fully time-multiplexed architecture is completely general and requires no pre-assumptions on data organisation. Due to this generality, a full time-multiplexed system can be built with only one type of data processing hardware.

By minimising the regional segmentation (and therefore maximising the time-multiplexing factor, P, accordingly), it is possible to overcome the problem of transferring data between processing nodes. As a single track may be associated with stubs that are created in multiple regions of the detector, each with a potentially separate readout path, it is crucial that these stubs are collected into a single processing node. As bandwidth is limited, time-multiplexing allows this to be accomplished in a *direct downstream* approach, without the additional complication and latency associated with *sideways communication*. A fully time-multiplexed system does not require data sharing or duplication over boundaries, therefore by definition it requires less data to be transmitted (but this does not necessarily mean fewer links) than any other design architecture, for a given data resolution.

Redundant nodes can be easily added to a time-multiplexed system. By providing spare readout links in the DTC, a redundant node can be switched in by software if a processing node fails. Alternatively, data from $1/P$ events can be duplicated into a spare node, allowing algorithm or firmware changes to be tested in the system (on real data), without affecting operation. In addition, in a fully time-multiplexed design, the failure of a single (time) node is equivalent to prescaling the trigger. This is in contrast to the failure of a single (regional) node in a conventional system, which would cause imbalances in topological triggers. Similarly, the desynchronization of a time node would not impact the remaining nodes.

As just one time node is required to demonstrate the entire system, a time-multiplexed design allows a verified node to be scaled up with confidence. This is different to a regionally segmented design where regional differences may mean different firmware, bandwidth requirements, and challenges. In addition, communication between regional sectors can add additional latency and complication, and should also be demonstrated properly. As the firmware in a time-multiplexed design is identical within each time-node, a time multiplexed design is likely to require fewer different firmware builds, and a simpler firmware (and software) management and deployment strategy than a regionally segmented trigger. It should therefore also be easier to make changes to the firmware without altering data paths.

3.2.2 Data Delivery and Regional Segmentation

By treating the DTC as the first layer in a time-multiplexed system, it should be feasible to stream the full set of stubs for a large fraction of the detector into a time node, or TFP. While it would be desirable to transfer data from the entire tracker into a single processing node, in practice this is limited by the number of input links and overall bandwidth into a single FPGA processor. In addition to time-multiplexing, it is therefore required to divide the tracker into nine (symmetric) regional sectors, henceforth called *nonants*, where it is ensured that all stub data required to reconstruct tracks within each nonant are contained within a single processing node. This works because the tracker modules are cabled to the DTC boards such that a given sub-set of ($N_{DTCs}/9$) DTC boards are guaranteed to receive data *only* from modules positioned in a single ϕ-nonant of the detector [5]. Until 2017, the proposed cabling scheme was based on a eight-fold symmetry, which naturally leads to a tracker segmentation into eight (symmetric) *octants* in ϕ. As a result of this, this thesis includes some results with an octant segmentation, where the flat barrel geometry is used. All results given for the tilted barrel geometry assume nonant divisions. Due to the nonant segmentation, one redundant time-node would require nine additional TFPs.

The proposed system architecture is depicted in Fig. 3.1. In order to handle duplication of data across hardware boundaries a simplification can be applied at the DTC-TFP interface. Defining processing nonant (octant) boundaries that divide the tracker into uniform 40 (45) degree ϕ-sectors, each rotated by approximately 20 (22.5) degrees in ϕ with respect to the cabling nonant (octant) boundaries, implies that a DTC handles data belonging to no more than two neighbouring processing sectors. The first step of the DTC is to unpack and convert the stubs from the front-end links to a global coordinate system. A globally formatted stub can be described adequately with 48 bits. This is followed by an assignment of every stub to one of the two regions, or if it is consistent with both, by duplicating the stub into both processing sectors. This duplication would occur whenever a stub could be consistent

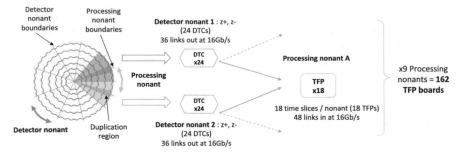

Fig. 3.1 Preliminary system architecture, whereby DTCs in two neighbouring detector nonants time-multiplex and duplicate stub data across processing nonant boundaries before transmission to the TFP boards. With 18 time nodes, and nine processing nonants, the full track finding system would consist of 162 TFPs. Duplication regions are defined in Chap. 4

with a charged particle in either processing sector, from the knowledge that a track with $p_T = 2$ or $3\,\text{GeV}$ defines a maximum possible track curvature. In addition, the measurement of the stub local bend can also be employed to minimise the fraction of stubs duplicated to both sectors. The exact logic is identical to that which will be described to assign stubs to sub-sectors in ϕ in Sect. 4.3.

As described, an advantage of a time-multiplexed design is the ability to demonstrate a full track-finder with a very small quantity of hardware; demonstration of one processing node is equivalent to demonstration of all processing nodes. This simplicity has allowed a TFP demonstrator to be built with currently available hardware. As this demonstrator is constructed with currently available hardware (with limited bandwidth capability), it is designed to process one sector (specifically one octant) of the tracker at a time, but with an extended time-multiplex period of 36, in comparison to the proposed final system which is envisioned to have a time-multiplex period of 18, and will be discussed in Sect. 7.1.

3.3   The Track Finder Demonstrator

3.3.1   Overview of Firmware Architecture

The proposed TFP and its demonstrator are divided into a number of individual steps and components, each of which are described and justified in the following chapters. Each of these steps can be operated standalone, or in a complete chain from TFP input to TFP output.

- Hough Transform (HT) - A highly parallelised first stage track finder that identifies groups of stubs that are coarsely consistent with a track hypothesis in the $r - \phi$ plane, reducing combinatorial background in the downstream steps. The Hough transform itself is preceded by a Hough Transform Preprocessor (HTP), which is responsible for organising the stub data into a convenient format and subdividing the stubs from the sector into finer sub-sectors in η and ϕ to simplify the track finding task and to increase parallelisation. The HT and its preprocessor are described in detail in Chap. 4.

- Track fitter - A track cleaning and precision fitting algorithm which acts on the set of track candidates to remove fake tracks, remove unwanted stubs from track candidates, and calculate fitted helix parameters. This thesis presents the option of a Kalman Filter (KF), as described in detail in Chap. 5.
- Duplicate Removal (DR) - A filter that uses the precise fit information to remove any duplicate tracks generated by the HT that remain following the track fit. Section 5.5.1 describes the proposed duplicate removal algorithm and implementation. In the demonstrator TFP the DR firmware is implemented within the same FPGA as the track fit.

These TFP algorithms and their order are shown in Fig. 3.2, and relate to the firmware components described in the following chapters. As the hardware is not yet available to fit all the required logic for a proposed TFP onto one processing card, it has instead been chosen to use several currently available FPGA processing cards to emulate the FPGA logic resources that may be available in a final TFP. To this end, eight MP7 boards (see Sect. 3.3.2.1) are currently used for the demonstrator chain. The TFP itself is implemented on five boards: one being used for the HTP, two for the HT, and two more for the KF (including the DR).

An additional board, the sink, is used to capture the track-finder output from up to thirty simulated physics events before being read out via the IPbus protocol [6].

Two boards, named sources, each represent data from a set of up to 36 DTCs. Each source board is implemented as a large buffer for the storage of stub data from a detector octant, where the data are loaded directly from simulation via the IPbus protocol. Each output stream from the source boards represents a separate DTC, injecting pre-formatted 48-bit stubs into the HTP, and is capable of playing

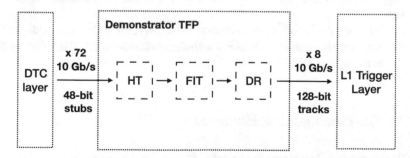

Fig. 3.2 A diagrammatic overview of the TFP algorithms, which are each described in detail in the following chapters. The connectivity between the demonstrator TFP, and the proposed DTC and L1 trigger layers is shown. In this diagram, each box (bordered by dashed lines) does not represent an individual FPGA or processing card, but instead a step in the TFP algorithm. All steps within the center box (bordered by dotted lines) are part of the demonstrator TFP

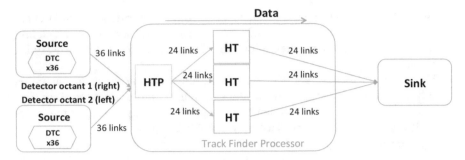

Fig. 3.3 An overview of the Stage 1 TFP demonstrator, illustrating the main logic components and their interconnectivity, each described in detail in the following sections. This preliminary stage was used to test and debug the HT firmware prior to the development of the track fitting implementations

up to thirty consecutive events through the demonstrator. Two sources are required to emulate how data from two adjacent sectors can feed a single TFP with tracks that cross the sector boundary.

The TFP demonstrator assembled in stages. Stage 1 of the demonstrator (completed May 2016) consisted of two source boards, one HTP board, three HT boards (operating in parallel), and one sink board, optically connected as shown in Fig. 3.3. This setup was used to demonstrate and test the rate reduction capabilities of the HT prior to the development of the track fitting firmware. For the ultimate, Stage 2 demonstrator (completed December 2016), one fewer MP7 board was used for the HT, and two additional boards were added (in parallel) between the HT and the sink, for the track fitting and duplicate removal layer. An additional layer of two boards (in parallel) was connected between the HT and the track fitting. These boards were connected to allow testing of larger or unoptimised track candidate processing firmware implementations. For the *full chain* demonstration presented in this thesis, these boards were configured to run a null algorithm. The connections between boards in the Stage 2 demonstrator are illustrated in Fig. 3.4, where each box corresponds to one MP7.

For standalone testing of firmware blocks, or parallel data taking alongside the full chain, an additional board is also installed in the demonstrator crate. The demonstrator crate is shown in Fig. 3.5.

3.3.2   The Demonstrator Hardware

The TFP hardware demonstrator consists of a number of daisy chained MP7 processing cards.

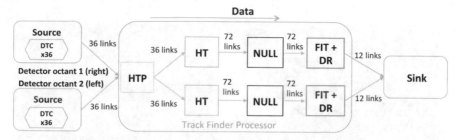

Fig. 3.4 An overview of TFP demonstrator (during Stage 2 operation), illustrating the main logic components and their interconnectivity, each described in detail in the following sections. Each box represents one MP7. The TFP demonstrator is capable of processing data from up to 72 input links (one per DTC). This allows some margin in the exact number of DTCs, which is yet to be determined

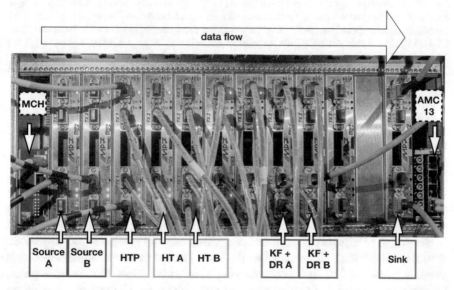

Fig. 3.5 A photograph of the demonstrator crate at Stage 2 operation. It is equipped with eleven MP7-XE boards, an AMC13, MCH and the required optics, as shown in Fig. 3.4. The TFP algorithms implemented in each board are labelled. The direction of data flow is from left to right

3.3.2.1 The MP7 Processing Card

The MP7 card [7], developed by Imperial College London, was designed to be a generic data stream processor, in the MicroTCA format [8–10]. As a consequence of this requirement, it is equipped with a fully symmetric, all-optical data interface and a large monolithic FPGA. A photograph of the top (right hand side) and bottom (left hand side) surfaces of the card is shown in Fig. 3.6. The chosen FPGA, the V7-690 [11], provides 80 serial links in each direction, which each run at up to 13.1 Gb/s (for the '−3' speed grade variety). In practice, eight of these links are dedicated to

Fig. 3.6 Photograph of the MP7 processing card [7]. The top side of the card is shown on the right hand side, and the bottom side of the card is shown on the left hand side. The optics and heat sinks are not mounted in these images. Some of the important components are labelled as follows: (1) the V7-690 FPGA; (2) six Avago MiniPOD transceivers; (3) six Avago miniPOD receivers; (4) the Xilinx CoolRunner-II XC2C256 CPLD [13], used for JTAG and boot management; (5) the USB-2 interface; (6) the microSD card interface; (7) the ATMEL microcontroller; (8) the MicroTCA backplane connector; (9) two 72 Mbit QDR-II SRAM chips; (10) CPLD and JTAG interface

backplane connections such as a PCIe, a SATA, and a Gigabit Ethernet line. The V7-690 also contains 3600 DSPs, 1470 Block RAMS, 433,000 LUTs and 866,000 Flip Flop registers. This chip was selected due to its high resource density, which fulfils the requirements of the CMS Phase I trigger [12], where complex triggering algorithms must be executed within a few μs. The high bandwidth and logic resource count of this chip, and the flexibility of the card make the MP7 ideal for demonstrating the Phase II track finder with currently available technology.

Six Avago MiniPOD optical interface transmitters and receivers [14] (for a total of twelve MiniPODs), are mounted on the MP7. Each of these transmitters/receivers provides twelve optical links, each running at up to 10.3 Gb/s. On the front panel, four standard 48-way MTP connectors are mounted, of which only 36 channels are used per connector. Two MTPs are dedicated to transmitting, and two to receiving data.

Up to 2×72 Mb of fast (550 MHz DDR) static RAM is also mounted on the card. In addition an Atmel AT32-UC3A-3256 microcontroller [15] is used for IPMI communications and monitoring, a USB-2 interface, and a microSD card interface.

The MP7 dissipates about 75 W when under load, of which approximately 35 W is dissipated by the FPGA [16].

Infrastructure tools were developed for the MP7 [17], including core firmware to manage transceiver serialisation/deserialisation, data buffering, I/O formatting, board and clock configuration as well as external communication via the Gigabit Ethernet interface. A diagram of the MP7 infrastructure firmware is shown in Fig. 3.7. Any algorithms deployed are segregated from the firmware responsible for these tasks, allowing a system such as the TFP demonstrator to be built up of processing

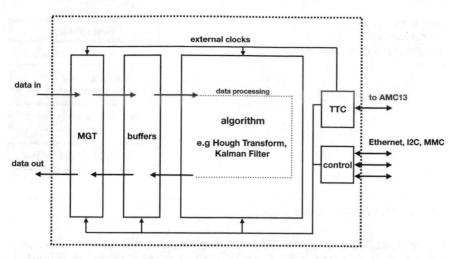

Fig. 3.7 Diagrammatic representation of the MP7 infrastructure, as used in the track finder demonstrator. Additional components of the infrastructure firmware such as the DAQ and readout control are disabled for the track finder demonstrator, as they are unused but take up a significant fraction of FPGA resources. Data enter the algorithm via MGTs. External (such as those coming from the AMC13), and internal clocks are supported

blocks, each running on a single MP7, daisy-chained together with high-speed optical fibres. Division of the demonstrator in this way allows firmware responsibilities to be easily divided between personnel, provided I/O formats between the processing blocks are defined. Parallelising or daisy-chaining algorithms across multiple boards allows estimation of final system performance, without limitations from the resources available in currently available technology. In addition, an upper limit on the total FPGA logic requirements for a future processing card can be extracted from the demonstrator, as detailed in Sect. 7.1.

The control software used for the MP7 demonstrator uses IPbus, a MicroHAL protocol in a C++ Python wrapper, as shown in Fig. 3.8. The Python interface is known as the MP7Butler, and the track trigger control software (also in Python) inherits many of the MP7Butler methods, with some additional custom classes. With this software, a simple command line interface can be used to upload new firmware images, modify settings (via registers) within the firmware, upload user-created data into buffers, and play data patterns through the firmware. It can also be used to print and parse readout data that comes over IPbus/MicroHAL.

The specific variant of MP7 used for the demonstrator is the MP7-XE, which is equipped with an FPGA specified to a '−3' speed grade. This speed grade is required for some of the more complex and large track finder firmware implementations to run at 240 MHz.

Fig. 3.8 Diagrammatic
representation of the track
trigger demonstrator control
software

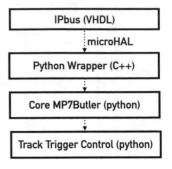

3.3.2.2   The Demonstrator Apparatus

The demonstrator is located at the TIF, a large hall at CERN dedicated to testing
and commissioning CMS tracker electronics. The demonstrator consists of one dual-
star MicroTCA crate [8], with vertical air cooling (shown in Fig. 3.5). Dual-star is a
backplane topology whereby two dedicated hub slots each connect to all node slots.
This is used for redundancy in the control hub modules. To provide the space needed
for large and power hungry (35 × 35 mm, order of 50 W) FPGAs, double width AMC
cards are required. Whereas typical dual width AMC crates contain 14 node slots,
two hub slots, and two power module slots; the crate used for the demonstrator has
been customised for the high power requirements (up to 1 kW per crate) of the CMS
trigger [18] to contain 12 node slots, two hub slots, and four power module slots.
This allows extra and redundant power when compared to the typical design. Each
power module can provide up to 800 W, is rated to 65 °C, and operates with 91%
efficiency. The power modules are manufactured by Vadatech [19]. The demonstrator
draws up to 500 W from each of two power modules, with the additional two used
for redundancy. Whereas typical MicroTCA crates allow one RTM per node, the
customised crate allows only six, none of which are used in the demonstrator. The
crate is manufactured by Schroff [20].

A standard control hub module, the (single width) MCH, manufactured by
N.A.T [21] provides Gigabit Ethernet communication as well as other standard I/O.
Typically the second hub slot is used for a redundant standard control hub module,
however in the demonstrator (and the CMS trigger) one hub slot is used for CMS
specific functions, and is occupied by a specific single width auxiliary card known
as the AMC13 [22] that distributes a common 40 MHz clock, and fast controls, to
each node card. It is equipped with a Kintex-7 FPGA [11], and is 10 Gb/s compatible
both on the backplane, and via 1–3 optical fibres on the front panel. The AMC13 is
also capable of collecting full DAQ data from the node cards, but this is not used in
the demonstrator. The use of the second hub slot for the AMC13 removes hub slot
redundancy from the system.

The TFP algorithms are implemented on a set of five MP7-XE cards. Figure 3.5 is a
photograph of the demonstrator crate, with the algorithms implemented in each MP7
labelled. The crate is installed in a standard CERN rack, which has been equipped

with a turbine, water cooling heat exchangers and an air deflector to ensure sufficient removal of heat. The rack also provides 3-phase power. A Dell PowerEdge R620 rack PC is connected to the MCH over Ethernet, and is used to control the demonstrator, inject and retrieve data, and run the demonstrator scripts and DQM software.

References

1. CMS Collaboration (2015) Technical proposal for the Phase-II upgrade of the CMS detector, Jun 2015, technical report CERN-LHCC-2015-010
2. Hall G (2016) A time-multiplexed track-trigger for the CMS HL-LHC upgrade. Nucl Inst Meth A 824:292–295. https://doi.org/10.1016/j.nima.2015.09.075
3. Amstutz C et al (2016) An FPGA-based track finder for the L1 trigger of the CMS experiment at the High Luminosity LHC, Jun 2016, IEEE-NPSS Real Time Conference, pp 1–9. https://doi.org/10.1109/RTC.2016.7543102
4. CMS Collaboration (2013) CMS technical design report for the Level-1 trigger upgrade, Jun 2013, technical report CERN-LHCC-2013-011
5. CMS Collaboration (2017) The Phase-2 upgrade of the CMS tracker, Jul 2017, technical report CERN-LHCC-2017-009
6. Ghabrous Larrea C et al (2015) IPbus: a flexible Ethernet-based control system for xTCA hardware. JINST 10: C02019. https://doi.org/10.1088/1748-0221/10/02/C02019
7. Compton K et al (2012) The MP7 and CTP-6: multi-hundred Gbps processing boards for calorimeter trigger upgrades at CMS. JINST 7:C12024. https://doi.org/10.1088/1748-0221/7/12/C12024
8. PICMG (2006) Micro Telecommunications Computing Architecture Short Form Specification, Sep 2006. https://www.picmg.org/wp-content/uploads/MicroTCA_Short_Form_Sept_2006.pdf
9. PICMG (2006) Advanced Mezzanine Card Short Form Specification, Dec 2006. http://www.picmg.org/pdf/AMC.0_R2.0_Short_Form.pdf
10. Di Cosmo M et al (2015) MicroTCA and AdvancedTCA equipment evaluation and customization for LHC experiments. JINST 10:C01008. https://doi.org/10.1088/1748-0221/10/01/C01008
11. Xilinx Inc (2017) 7 series FPGAs data sheet: overview, Aug 2017, product specification, DS180 (v2.5)
12. Zabi A et al (2017) The CMS Level-1 trigger system for LHC Run II. JINST 12:C01065. https://doi.org/10.1088/1748-0221/12/01/C01065
13. Xilinx (2008) CoolRunner-II CPLD family product specification, Sep 2008, DS090 (v3.1) https://www.xilinx.com/support/documentation/data_sheets/ds090.pdf
14. Avago Technologies (2013) MiniPOD AFBR-811VxyZ, AFBR-821VxyZ 10 Gbps/channel twelve Channel, parallel optics modules product brief, Mar 2013. https://docs.broadcom.com/docs/AV02-2842EN
15. ATMEL Corporation (2012) 32-bit AVR Microcontroller Summary, 2012, 32072SH-A VR32-10/2012. http://ww1.microchip.com/downloads/en/DeviceDoc/32072s.pdf
16. Iles G, Jones J, Rose A (2013) Experience powering Xilinx Virtex-7 FPGAs. JINST 8:C12037. https://doi.org/10.1088/1748-0221/8/12/C12037
17. CACTUS: Code Archive for CMS Trigger Upgrades. Accessed Feb 2018. https://svnweb.cern.ch/trac/cactus/wiki
18. Iles G, Hansen M, Gorski T, Hazen E (2011) CMS MicroTCA crate concepts & AMC card requirements, Jan 2011, ver. 0.9. http://joule.bu.edu/~hazen/CMS/AMC13/The%20CMS%20uTCA%20Crate%20v0.9_esh.pdf

19. VadaTech Inc (2014) MicroTCA overview a brief introduction to Micro Telecommunications Computing Architecture concepts, Mar 2014, ver 1.1. https://www.vadatech.com/media/article_MicroTCA_Overview.pdf
20. Pentair plc (2017) 14U 14-slot ATCA shelf user's manual, May 2017, Doc-No: 63972-344_R1.0.   https://schroff.pentair.com/wcsstore/ExtendedSitesCatalogAssetStore/Attachment/SchroffAttachments/Documents/63972-344.pdf
21. NAT GmbH, N.A.T-MCH user's manual ver. 1.31, Aug 2016, N.A.T GmbH Konrad-Zuse-Platz 9, 53227 Bonn-Oberkassel. http://www.nateurope.com/manuals/nat_mch_man_usr.pdf
22. Hazen E et al (2013) The AMC13XG: a new generation clock, timing, DAQ module for CMS MicroTCA. JINST 8:C12036. https://doi.org/10.1088/1748-0221/8/12/C12036

Chapter 4
The Hough Transform

4.1 The Hough Transform Algorithm

The classical HT is widely used to detect parametrically described curves such as circles or lines in an image that can contain noise or partial occlusion. Following the invention of the linear HT for machine analysis of bubble chamber photographs [1], the HT was generalised to identify positions of arbitrary shapes [2]. This chapter describes how it is possible to use the linear two dimensional HT to find track candidates within the $r - \phi$ plane of the Phase II CMS Outer Tracker, in real time.

In Sect. 2.5.1 it was shown that it was possible to derive the relationship between beam incident angle and p_T using standard geometry shown in Fig. 2.6 and the radius of curvature, given as a function of p_T in Eq. 2.2. This is accurate only if the assumption is made that particles relevant to the L1 trigger originate at or close to the interaction point.

Using Fig. 4.1 to understand the relationship between α; ϕ_0, the ϕ direction (angle) of the track in the transverse plane at the origin of CMS; and the local ϕ coordinate of the stub, simply labelled ϕ, it is possible to show that the path of the particle in the transverse plane can be written as

$$\frac{r}{2R} = \sin(\phi - \phi_0). \tag{4.1}$$

This formula uses the same units as Eq. 2.2.

For tracks with $p_T > 2\,\text{GeV}$, the radius of curvature is large (at least 1.75 m) so the small angle approximation can be used, which gives

$$\frac{r}{2R} \approx \phi - \phi_0. \tag{4.2}$$

In practice ϕ is defined relative to the centre of each sector, so edge effects at 2π or 0 do not need to be considered. In the scenario where the particle is assumed not to interact further within the detector (e.g multiple scattering and bremsstrahlung are

© Springer Nature Switzerland AG 2019
T. O. James, *A Hardware Track-Trigger for CMS*, Springer Theses,
https://doi.org/10.1007/978-3-030-31934-2_4

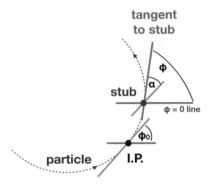

Fig. 4.1 Illustration of the angles involved in the derivation of the HT formula. The black dot represents the Interaction Point (I.P.), the red dot represents the position of the stub, the particle trajectory is shown by the dotted purple curve, the green line represents the tangent of the particle's trajectory at the stub position, and the yellow line represents the tangent of the particle's trajectory at the interaction point. Same coloured lines are parallel

ignored), one can assume that the (r, ϕ) coordinates of any stub produced by this particle must be compatible with this trajectory.

Combining this result with Eq. 2.2 one obtains the relation between the (r, ϕ) coordinates of a single stub, and a corresponding straight line in the Hough-space $(q/p_T, \phi_0)$,

$$\phi_0 = \phi - \left(\frac{Bcr}{2}\right) \frac{q}{p_T}. \tag{4.3}$$

It is now evident how one can transform a stub with coordinates (r, ϕ) to a line with intercept ϕ_0 and a gradient proportional to r.

Lines in Hough-space that correspond to the set of stubs produced by the same particle, will intersect at a single point (for now ignoring detector resolution effects). One can therefore apply this transformation to a set of stubs, and identify stubs associated with the same particle track from those that meet at a stub-line intersection. The intersection point also determines the $(q/p_T, \phi_0)$ track parameters. The process of transforming a set of stub locations in Cartesian space, to a set of lines (and a corresponding intersection point) in Hough-space is depicted in Fig. 4.2.

In this Hough-space, the gradient of each stub-line is proportional to the radius r of the stub, so is always positive. It is preferable to instead measure the radius of the stub using the variable $r_T = r + T$, where T is a chosen offset, to ensure that a given track consists of stubs with a suitably wide range of positive and negative gradients in $(q/p_T, \phi_0)$-space. This improves the precision with which one can measure the intercept point in a finitely granular array (which is necessary for implementation in firmware). This transforms Eq. 4.3 into

$$\phi_T = \phi - \left(\frac{Bcr_T}{2}\right) \frac{q}{p_T} \tag{4.4}$$

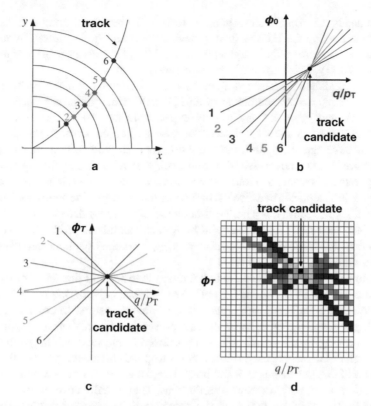

Fig. 4.2 Illustration of the HT. **a** shows the trajectory of a single particle and the stubs that it produces in the Cartesian $x - y$ plane. **b** shows the same six stubs as lines in $(q/p_T, \phi_0)$-space. Here each stub is represented by a straight line, and the point where the lines intercept both identifies a track, and determines its $(q/p_T, \phi_0)$ parameters. **c** shows the same six stubs in Hough space once subject to the transformation $r_T = r + T$. Figure **d** shows the digitised version of (**c**), where the Hough space is represented as an array. In **d**, the stub bend values are used to restrict the binning range in q/p_T

where the track parameters are now $(q/p_T, \phi_T)$, where ϕ_T is the track ϕ at a chosen radius T. In this new Hough-space, the stub-line gradient is proportional to r_T, so can be either positive or negative, as depicted in Fig. 4.2. The use of both negative and positive stub-line gradients improves the precision with which the intersection point can be measured, resulting in fewer mis-reconstructed or duplicate tracks.

To implement the HT algorithm in FPGA logic, it is necessary to subdivide the Hough-space into an array of cells, with an array range $|q/p_T| < q/p_T^{min}$ in one axis (where for our purposes $p_T^{min} = 2$ or 3 GeV), and $\phi_T^{min} < \phi_T < \phi_T^{max}$ in another axis. Stubs are binned into any cell that their stub-line passes through, and a cell with a certain number of hits can be marked as a track candidate. In general, the finer the array granularity, the more effective the HT (fewer combinatorial fakes). It has been found, however, that the cell size (or array granularity) could not be reduced beyond approximately 6.136 mrad, corresponding to 1024 ϕ_T bins for the entire 2π range of ϕ_T without making the HT sensitive to deviations from Eq. 4.4 caused by multiple scattering or detector effects. Typically around 32 (48) bins in q/p_T are used for $p_T^{min} = 3(2)$ GeV. This choice will be explained in Sect. 4.2.1.

A track candidate is identified if stubs from a minimum number of tracker barrel layers or endcap disks accumulate in an HT cell. Primary charged particles with $p_T > 2$ GeV and $|\eta| < 2.4$ are usually expected to traverse at least six of these stations. However, to allow for detector or readout inefficiencies, and for imperfect geometric coverage, the default threshold criteria used to identify a track candidate only require stubs in at least five different tracker barrel layers or endcap disks, and this requirement is reduced to four in the region $0.89 < |\eta| < 1.16$ to accommodate a small gap in acceptance between the barrel and the endcaps. The configuration of the number of layers required is a trade-off between allowing for detector inefficiencies, and an increased rate of combinatorial fake track candidates produced by the HT. Results that demonstrate the outcome of varying the radial offset, the cell width, and the layer criteria are given in Sect. 4.4.

Due to bandwidth limitations, it is not possible to feed all the data from a single event into one FPGA (or one HT array). As discussed in Sect. 3.2, the natural segmentation of the tracker is into either eight or nine ϕ sectors. In order to increase parallelisation, and therefore speed up data processing, it is also necessary to create virtual (meaning within the FPGA, and unrelated to physical cabling) sub-sectors within each FPGA, where each sub-sector corresponds to one array in HT firmware. Within an FPGA the sub-sectors can be as large or as small as needed to give optimum performance, within resource limitations. They usually cover a defined range of η and ϕ_T (for example, two ϕ_T and eighteen η divisions per ϕ-sector). The optimisation of the ranges of these sub-sectors is described in Sect. 4.3. They must be have wide enough overlapping regions to ensure that no tracks are lost due to subsector boundaries, but the required duplication must not be so large that bandwidth limitations in and out of each HT array are exceeded.

Each stub also contains its bend information, which can be used to estimate an allowed range in q/p_T of the particle that produced the stub. Using Eq. 2.3, one can derive an allowed range given by $(q/p_T)_{min} < (q/p_T) < (q/p_T)_{max}$, where

$$(q/p_T)_{max} = \frac{2}{crB}\alpha_{max} = \frac{2(b + k_b)\,\rho}{crB} \,,$$

$$(q/p_T)_{min} = \frac{2}{crB}\alpha_{min} = \frac{2(b - k_b)\,\rho}{crB} \,,$$

(4.5)

$\rho = (p/s)$ for barrel stubs and $\rho = (p/s) \cdot (z/r)$ for endcap stubs, and p and s are the pitch and separation of the two sensors in a module, respectively, b is the bend of the stub in units of strip pitch, and k_b is a configurable parameter such that the true bend is assumed to lie within k_b of the measured value. This will be discussed further in Sect. 4.3.1.

Each stub should therefore only be added to cells in the HT array within q/p_T columns compatible with this allowed range. This method is shown in Fig. 4.2. It substantially reduces the probability of producing combinatorial fake candidates, as demonstrated in the results that follow.

4.2 Firmware Implementations

Porting the Hough transform algorithm (which is relatively easily implemented in software), to an FPGA firmware design was a non-trivial task. The proposed firmware design evolved significantly over time, as new bottlenecks and optimisations were discovered.

4.2.1 Systolic Array Implementation

A systolic array architecture was initially proposed for the Hough transform firmware. A systolic array consists of a homogeneous network of data processing cells, each cell independently computing a partial result, as a function of data received from upstream cells. This result is then stored in each cell, and passed downstream. Systolic arrays are widely used for massively parallelized tasks, such as data sorting, or matrix multiplication. As the Hough transform requires a computation at each cell, a systolic array allows these computations to be executed in parallel, with no requirements to access any external buffers outside the cell.

The initial firmware task was therefore to design a Hough array cell, which could be multiplied many times over to fill a Hough array. Each cell would be connected to its nearest neighbours only. A block diagram representation of the systolic array firmware design is shown in Fig. 4.3. In this design, stubs that enter the HT are propagated to West, North, and South controllers which determine stub suitability for entrance to the array at the adjacent HT cells. This means no preprocessing or sorting of the data is needed prior to entering the HT firmware. The inner workings of each HT cell are depicted in Fig. 4.4, and each firmware cell corresponds exactly to a cell in the mathematical HT array in the $(q/p_T, \phi_T)$ plane. One column therefore corresponds to one bin in q/p_T, and one row corresponds to one bin in ϕ_T, respectively. Within the array, stubs propagate one column at a time from West to East, and have the ability to enter up to three possible rows. An entry check at the entrance to each cell determines if the stub is compatible with the cell coordinates, and if so, it is stored in one of three First In First Out (FIFO) memories, North West (NW), West (W), or South West (SW), depending on its entry position. An arbitrator takes one stub from these FIFOs per clock (choosing in order NW, W, SW), and sends the stub data to two locations in parallel. One path leads to a pair of DSPs that calculate the ϕ_T value (using Eq. 4.4) for that stub at the adjacent (next in line) q/p_T column. The other path leads to the BRAM, where the stub is stored. However, the bend of the stub is first

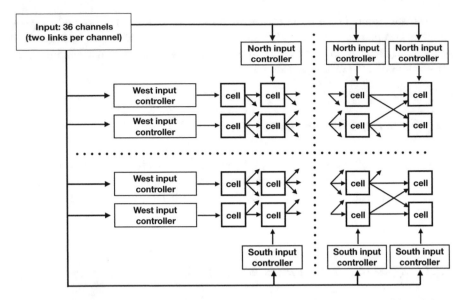

Fig. 4.3 The systolic Hough transform array, represented in block diagram format. Thirty-six input channels each propagate one stub per clock to all North, West, and South input controllers. If the stub is accepted for entry into the array at the specified input controller, it continues to propagate from cell to cell one column per clock, until it no longer meets the criteria

tested to ensure compatibility with the q/p_T location within the array (using Eq. 4.5), and only compatible stubs are stored. The systolic array cell can implement a check on the number of unique tracker layers/disks marked within the cell, or a check that a certain number of uniform bins in r are selected (up to a maximum of 16 bins). It is also possible to implement both of these criteria simultaneously which reduces the number of combinatorial fake track candidates by 41% when five sub-sectors in η are used, and by 16% when nine sub-sectors in η are used, in comparison to the second criterion alone (for events of $t\bar{t}$ and 200 pileup, flat barrel geometry, $T = 45$ cm).

Individual locations in the cell block memory exist for stubs associated with each HT sub-sector, meaning that only a single array must be implemented in firmware to process data associated with many sub-sectors at once. With a storage constraint of sixteen stubs per cell, per sub-sector, per event, the memory can be configured to separately bin stubs from up to 64 sub-sectors at once. This technique is massively more efficient that having to construct an entire systolic array for each sub-sector, and facilitates a natural load balancing of data, as fluctuations in local data rate are equalised over many sectors.

It should be noted that in this implementation, the q/p_T and ϕ_T values that each column and row represent are entirely configurable during operation of the firmware in an FPGA (even to non-linear distributions of values). This allows maximum flexibility in the range of the Hough-space (and indeed physical space) that will be processed at any one time. An additional feature is the ability to encode error boundaries

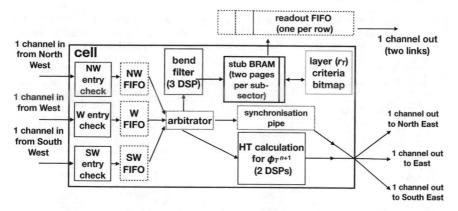

Fig. 4.4 The systolic array Hough transform cell, represented in block diagram format. Stub data flows in the direction of the arrows. Each cell has three possible inputs, but can only process one stub per clock. Stubs that pass all selection criteria are stored in an appropriate location in BRAM (corresponding to their sub-sector ID) and are sent to the three nearest neighbour cells in the next column

to the cell coordinates, in order to account for rounding errors in the cell calculations. As a fully integer-based calculation was later adopted, this feature was unused, but could be reinstated if overlapping cells were required to account for tracking uncertainty (such as multiple scattering of low p_T particles), at the expense of additional track candidate duplication.

At the end of a single event (BX) worth of data, a dummy stub is sent to the systolic array, which propagates to all cells. This dummy stub contains a bit (that is usually unused) to trigger the readout of the array. Each row is connected to a single readout FIFO, which can be connected either to serial links or a single large readout buffer.

After including the MP7 infrastructure firmware, and the area constraints associated with it, a systolic array of size 12 columns, and 14 rows was able to route, build, and meet timing constraints at 240 MHz within the V7-690 FPGA. As shown in Table 4.1, the systolic array and its input controllers utilised approximately 20% of the resources available within the chip each. Further increases in the size of the array were limited by routing constraints (not by resource usage), in particular in the fan-out to all input controllers. Although systolic arrays of size 14×14 and 18×9 could be placed and routed within the chip, the high resource utilisation meant that timing constraints were not able to be met for some paths within the cell logic, and for many of the input lines to the North and South input controllers. Due to the black box nature of optimisation algorithms employed by the firmware placing and routing tools, trial and error was used to find the maximum possible array size in the V7-690.

As the systolic array memory can sort and store stubs from up to 64 ϕ sectors, the array can achieve the required granularity in ϕ_T with 14 rows in ϕ_T (resolving to 896 bins in ϕ_T for the entire tracker). A limiting factor with respect to the array dimensions is that, as cells can only talk to their three nearest neighbours in the

Table 4.1 Resource utilisation of a 14 row, 9 column systolic array, as implemented in the V7-690 FPGA [3]. Raw numbers are given, in addition to numbers as a percentage of the available resources in the V7-690. Numbers for the array, the input controllers, and totals are listed individually, as well as accumulated. A description of the resource units can be found in Sect. 1.4

Firmware	LUT	DSP	FF	BRAM
1 array cell	582	5	1182	5
1 input controller (S/N)	2350	3	700	0
1 input controller (W)	2350	2	700	0
Systolic array	73,000	630	149,000	630
All input controllers	75,000	76	22,000	0
MP7 infra.	96,000	0	89,000	330
Total	244,000	706	260,000	960
Total (no infra.)	148,000	706	171,000	630
Available in V7-690	433,000	3600	866,000	1470
Fraction of V7-690 [%]				
Systolic array	17	18	17	43
All input controllers	17	2	26	0
MP7 infra.	22	0	10	22
Total	56	20	53	65
Total (no infra.)	34	20	43	43

adjacent columns, the cell width in q/p_T cannot be so large that a stub may have the coordinates to be binned in more than three adjacent ϕ_T bins within one q/p_T bin. This is illustrated in Fig. 4.5. Due to this limitation, it is required that for this ϕ_T granularity, the arrays must have no fewer than 25 q/p_T bins if the radial offset is set to $T = 65$ cm, and no fewer than 36 q/p_T bins if the radial offset is set to $T = 45$ cm. As a result, one would be required to use two or three V7-690 FPGAs in order to process the data from one TFP sector. (This problem could be eliminated if cells were wired to five nearest neighbours instead of just three, but this is challenging considering the density of resource usage already in the implementation). Splitting the array in q/p_T across two or three FPGAs is possible as long as the stub data can be routed to all HT processors within bandwidth limitations, using the stub bend to constrain the stub to HTs that cover a compatible range of q/p_T. As the resolution of the p_T estimates based on the bend is quite poor, significant additional duplication would still be required.

The minimum First In First Out (FIFO) latency (not to be confused with FIFO memory buffers) of the systolic array is given by the time taken for all stubs to enter and traverse the array (including the end of event marker stub). In practice, this time is bounded by the chosen time multiplex period, P. Simulations with datasets at 140 pileup, however, confirm that there will almost always be stub traffic bottlenecks. A major issue with the systolic array design, therefore, is that the heavily event dependent data flow (stub traffic) may lead to problems with bottlenecks where some

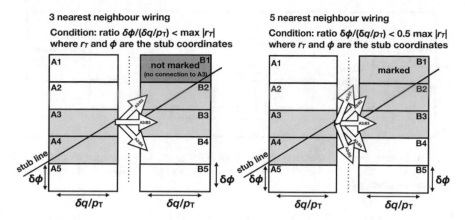

Fig. 4.5 Illustration of the HT array gradient constraint, in scenarios where each cell is able to talk to three or five nearest neighbours in the adjacent column

array cells are idling while stubs upstream are waiting to be processed. This data flow is difficult to predict and simulate, as it depends on exact number, coordinates, ordering, and timing of all stubs entering the array. For example, in a certain event there may be a heavy bottleneck in one input controller, but this scenario could be very rare and specific, and is therefore not observed within a typical test dataset.

The complexity and interconnected nature of the systolic array means that it may not be well suited to a track finding processor where scalability is highly valued. While some aspects of the systolic array have potential to be simplified, such as the heavy use of DSPs that could be reduced by using addition rather than multiplication (or sequestered to a preprocessor block), routing complications will remain. As each input channel must be connected to each input controller, and subsequently to every cell within the array, it becomes very difficult to scale up the array size without difficulties placing and routing the firmware implementation.

4.2.2 Pipelined Implementation

As a consequence of the challenges associated with the implementation of the systolic array, an alternative proposal for the HT firmware design has been investigated. The *pipelined* implementation eliminates the concept of a fully systolic structure in favour of a fully pipelined approach. This approach separates the work of the stub sorting/routing; and the HT array binning and track candidate selection. To do this, it assumes the existence of a Hough Transform Preprocessor that would route the stubs to their required virtual sub-sectors (in η and ϕ). Stubs associated with each of these sub-sectors are routed to individual Hough arrays, one for each sub-sector. The firmware component that represents one HT array per sub-sector is known as the HT *segment*, and is shown in Fig. 4.6. Each segment is independent, and has two input

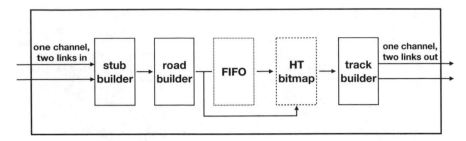

Fig. 4.6 Firmware implementation of one pipelined HT segment. One segment is needed for each HT sub-sector. Each pipelined segment processes one stub per clock. The function of each firmware block is described in the text

and two output links, which can transfer one 64-bit stub per 240 MHz clock. These segments are fully pipelined, and each one contains a 32×32 map in distributed memory, corresponding to the Hough transform array.

The HT map within the segment represents the HT array. This is depicted in Fig. 4.7. Each location in this map consists simply of a 6-bit distributed memory, one bit for each tracker layer or disk that may be crossed by the track. The precise mapping of bit to layer/disk is a function of η, where at high η some disks are combined so that they can be represented with a single bit. These bits make it straightforward for the candidate cells, with stubs in a minimum number of layers/disks, to be selected. The block labelled road builder calculates the appropriate row numbers for each stub (the values for each column are calculated in parallel). The FIFO memory stores the stubs and their roads. When the layer/disk criterion is met for a certain HT cell, the stubs associated with that cell are read out as track candidates by the track builder.

An advantage of this pipelined structure is that it negates the need for traffic management, and it is guaranteed that each array will process one stub per clock, if a stub is available. For this reason, the latency is much easier to predict in comparison to the systolic approach. The FIFO latency in clock cycles is given by the number of stubs that enter the segment plus the number of stubs on track candidates that must be read out. As each segment is a separate firmware block, scaling to more or fewer segments (each representing one array for one sub-sector) is straightforward. This is an advantage over the systolic approach, where routing and timing difficulties occur when the size of the systolic array is scaled up beyond about 14×12.

Compared to the more flexible systolic array cell, the simplicity of the pipelined segment makes it more difficult to add additional functionality, such as new selection criteria, without a design overhaul and hugely increased resource usage. The design of the pipelined segment also makes it difficult to enable a stub to mark multiple rows in ϕ_T within a single q/p_T bin. To overcome this limitation, which has a significant impact on the efficiency of track finding, two maps are required per segment, which equates to a doubling of memory usage when compared to the initial concept.

Eight pipelined segments, each representing a 32×32 HT array are able to place, route, and run at 240 MHz in the V7-690 FPGA. The resource utilisation for

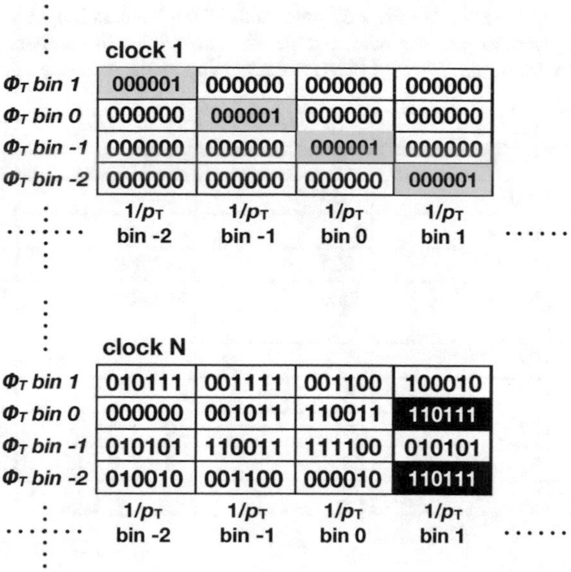

clock 1			
Φ_T bin 1	000001	000000	000000	000000
Φ_T bin 0	000000	000001	000000	000000
Φ_T bin -1	000000	000000	000001	000000
Φ_T bin -2	000000	000000	000000	000001
	$1/p_T$ bin -2	$1/p_T$ bin -1	$1/p_T$ bin 0	$1/p_T$ bin 1

clock N			
Φ_T bin 1	010111	001111	001100	100010
Φ_T bin 0	000000	001011	110011	110111
Φ_T bin -1	010101	110011	111100	010101
Φ_T bin -2	010010	001100	000010	110111
	$1/p_T$ bin -2	$1/p_T$ bin -1	$1/p_T$ bin 0	$1/p_T$ bin 1

Fig. 4.7 Illustration of the track candidate bitmap for a small number of HT cells in the pipelined HT implementation, at clock 1 in the event processing (upper), and at an arbitrary time later, clock N (lower). Each cell in this map corresponds to one cell in the HT array. Each bit in the memory corresponds to one of six tracking layers. These bits are initially set to 0 at the start of every event, and are set to 1 when a stub matches the cell criteria and is located in the corresponding tracker layer. Grey cells are those marked by the first stub (in layer six) to be processed by the array. The black cells are those that correspond to a track candidate at clock N

this design is given in Table 4.2. Although these eight segments only use about 22% of the available LUTs with the chip, timing issues made further scaling up of the design challenging. As the pipelined segment is large, and not easily factorisable, it is difficult to maximally utilise (fill up) the chip. The resource usage of the HTP must be taken into account when comparing this implementation with the systolic. As a minimum of 36 segments (each corresponding to one of 36 sub-sectors) are required to demonstrate one TFP, using this design four to five MP7 boards (plus an additional one if including the HT preprocessor) would be required in the hardware demonstrator for the HT algorithm alone (excluding track fitting, sources, sinks, and duplicate removal). This should be compared to two to three MP7s for the systolic implementation.

4.2.3 Daisy Chain Implementation

A new design was developed to overcome the resource usage issue with the fully pipelined array [4]. This design utilises a mostly pipelined data flow structure, with

Table 4.2 Resource utilisation of a 32×32 pipelined HT implementation, in a V7-690 FPGA [3]. Raw numbers are given for each segment, and the total, in addition to numbers as a percentage of the available resources in the V7-690 FPGA. A description of the resource units can be found in Sect. 1.4

Firmware	LUT	DSP	FF	BRAM
1 segment	11,875	128	8750	15
8 segments	95,000	1024	70,000	120
MP7 infra.	96,000	0	89,000	330
Total	191,000	1024	169,000	450
Available in V7-690	433,000	3600	866,000	1470
Fraction of V7-690 [%]				
8 segments	22	28	8	8
MP7 infra.	22	0	10	22
Total	44	28	18	30
Total (no infra.)	22	28	8	8

Table 4.3 Baseline HT configuration for the flat and the tilted barrel geometry, including the number of sub-sectors, and bins in the array. These flat barrel configuration settings are used in the MP7-based TFP demonstrator. Unless stated otherwise, flat barrel emulation results also use these settings. Results shown in this thesis for the tilted barrel geometry represent an emulation with these settings (unless stated otherwise)

	Flat barrel	Tilted barrel
# ϕ sectors	8	9
# ϕ sub-sectors	2	2
# η sub-sectors	18	18
# bins in q/p_T per array	32	32
# bins in ϕ_T per array	64	64
Radial offset, T [cm]	58	61.2

individual I/O for each Hough array, but it eliminates the need for a large map of bits for each array, significantly reducing the resource utilisation. Rather than duplicating the entire array map to allow up to two ϕ_T values per bin, the daisy chain implementation only duplicates stubs that require two ϕ_T values. This sacrifices the truly pipelined nature of the design, but allows a vastly reduced resource usage when compared to the pipelined (or systolic) implementation. In addition, this design re-optimises memory usage, and changes to utilise block RAM in place of distributed RAM whenever possible. Block RAM is more compact and therefore allows easier placement and routing. Henceforth all discussion of the HT firmware, emulation, and associated results will refer to this *daisy chain* implementation with the configuration shown in Table 4.3.

The daisy chain HT has been implemented in firmware, with the assumption that the demonstrator TFP will employ 36 HT arrays running in parallel (equivalent to $2\phi \times 18\eta$ sub-sectors). Similarly to the pipelined implementation, each individual HT array processes data from one input channel (two links), corresponding to the stubs consistent with a single sub-sector, as sorted by the HTP.

The design of each daisy chain array can be split into two stages: the filling of the array with stubs; and the readout of the track candidates that it finds. Each stage processes one stub at 240 MHz. On average, when reading simulated $t\bar{t}$ events at 200 pileup (flat geometry), each HT array must process 93 stubs per event.

The firmware design of each independent HT array is shown in Fig. 4.8. It consists of 32 firmware blocks labelled *columns*, each corresponding to one of the q/p_T columns in the HT array, and a number of firmware blocks named *book keepers*, each responsible for managing a subset of columns. The default configuration of columns and book keepers is shown in Fig. 4.8. There are twelve book keepers, each of which communicates with between two and three daisy-chained columns.

It is also possible to use a single book keeper, connected to all 32 daisy-chained columns. This is shown in Fig. 4.9. However, this configuration limits the rate at which reconstructed tracks can be output from the HT, as the book keeper can only output one stub per clock cycle, and therefore about 216 stubs per event (before the next event must be read out). Because jets can generate a high track occupancy within a single sub-sector, significant data truncation was observed due to this design flaw. In $t\bar{t}$ events with 200 pileup, an overall track finding efficiency loss of about 6% was observed. This problem is resolved by splitting the daisy-chain into twelve sub-chains, each of whose track candidates is read out in parallel via an independent

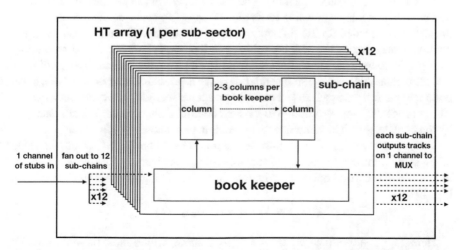

Fig. 4.8 Firmware implementation of one HT array in the daisy chain implementation, as used within an individual sub-sector. In each of the twelve pages visible in the figure, a book keeper is connected to a daisy chain of two to three columns. Eight book keepers × three columns and 4 book keepers × two columns = 32 columns in total. Internal components are shown as boxes and data paths as lines, where arrows indicate the direction of data flow

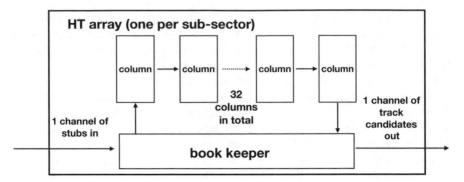

Fig. 4.9 Firmware implementation of one HT array in the simplified daisy chain implementation, as used within an individual sub-sector. In this implementation each array consists of 32 columns connected to one book keeper. Internal components are shown as boxes and data paths as lines, where arrows indicate the direction of data flow

book keeper. In this design, each output channel is fed by the multiplexed output from six book keepers, where each of these six book keepers represents an array from a different sub-sector, making it unlikely that a single jet could cause a high data rate in all six. The multiplexing reduces the total number of output channels from all the HT arrays in an octant to 72. To improve load balancing further, adjacent columns connected to the same book keeper correspond to non-adjacent q/p_T columns, such that each daisy-chain processes data associated with both low and high p_T columns.

The book keeper receives one stub per clock cycle from the input channel, which it stores within a 36 Kb block memory. The book keeper then sends the stub data to the first column that it is responsible for in the HT array. However, as the z coordinate of the stub is not needed for the HT, only a subset of the stub information is sent to the column, consisting of the stub coordinates in the (r_T, ϕ_T) with reduced resolution, an identifier to indicate which tracker layer the stub is in, the range of q/p_T columns that are compatible with the stub bend (this range is pre-calculated in the HTP), and a pointer to the full stub data stored in the book keeper memory. On each clock cycle a stub propagates from one column to the next column along the daisy chain managed by the book keeper. The components of a column are shown in Fig. 4.10.

The stub propagation from column to column is based on Eq. 4.4, where the value of ϕ_T at the right-hand boundary of the nth column is given by the following calculation, which is carried out in the component labelled Hough transform calculator in Fig. 4.10.

$$\phi_T(n) = \phi_T(0) + n \cdot \Delta_{q/p_T} \cdot r_T \ . \tag{4.6}$$

Here, Δ_{q/p_T} is the fixed width of a q/p_T column, which must be multiplied by an integer n, defining the q/p_T column index. The value $\phi_T(0)$ is given by the ϕ coordinate of the stub. To simplify the firmware, both $\phi_T(n)$ and ϕ are measured relative to the azimuthal angle of the centre of the sub-sector. Furthermore, the constants appearing in Eq. 4.4, such as the magnetic field, are absorbed into the

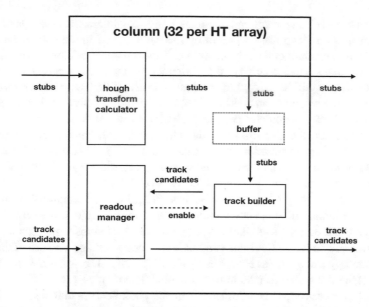

Fig. 4.10 Firmware implementation of one column in the daisy-chained Hough transform implementation, which corresponds to a single q/p_T column in the HT array. A number of columns are daisy-chained together, starting and ending with the book keeper

definition of q/p_T. To minimise the number of calculations within the array, the r_T values are expressed by the HTP in units of the width of a ϕ_T column divided by the width of a q/p_T column.

Since the range of q/p_T columns that are compatible with the stub bend is precalculated, only a comparison is needed to check column compatibility with the bend. Two DSPs are required to carry out the HT calculation described in Eq. 4.6, since the $\phi_T(n)$ values of both the left and right boundaries of the column are needed for the next step.

In each q/p_T column, the array has 64 ϕ_T cells. Stubs with a steep stub-line gradient can cross more than one (but by construction, never more than two, as described in Sect. 4.2.1) of these cells within a single column. Such cases are identified by comparison of the values of ϕ_T, from the HT calculation, at the left and right boundaries of the column. If a stub is consistent with two cells in the column, then it must be duplicated and stored within the buffer. The second entry will then be processed at the next available gap in the data stream. The track builder block places each stub it receives into the appropriate ϕ_T cell, where it implements the 64 ϕ_T cells using a segmented BRAM. Each track builder uses two 18 Kb block memory, organised as two sets of 64 pages, where the two sets take it in turn to process data from alternate time-multiplexed events. Each page corresponds to a single ϕ_T cell and has the capacity to store up to 16 stub pointers (the maximum number of stubs that can be associated with a track candidate).

For each ϕ_T cell in the column, the track builder maintains two records of which barrel layers or endcap disks were hit by the stubs stored in the cell. Each record corresponds to one half of the sub-sector in η. If the threshold criterion on the number of hit layers/disks is met in either of these two records, then a track candidate has been found, so the cell will be marked for readout. The use of half sub-sector information provides the equivalent to an additional factor of two in η segmentation in terms of the number of track candidates per event, obtained without the cost of doubling the parallelisation, and therefore logic. On the other hand, the fraction of correct stubs on the track candidates is not improved by this method as, due to firmware limitations, all the stubs stored in cells meeting the threshold criteria are read out, rather than only those compatible with just one of the sub-sector halves.

The readout manager is responsible for shifting the track candidate stubs from column to column, until there are no more stubs in the pipeline. It then enables read out of the track builder, such that a contiguous block of stubs from matched track candidates will be created. A track candidate stub now contains a record of the track parameters, ϕ_T and q/p_T (as Hough array indices), and a stub pointer, which is used to extract the full stub information from the book keeper memory.

A multiplexer groups the candidates from six book keepers onto a single output, resulting in a total of 72 outputs from the HT per TFP. At this stage a full 36 channel to 36 channel load balancing (using a configuration of the same routing firmware used for the HTP router, as described in Sect. 4.3.1.1) is applied across sub-sectors so that if an excessive number of tracks is found in a single HT array, typically within dense jets, candidates are assigned to different outputs to ensure all data is passed on to the next stage efficiently. This is required for downstream track fitters to process the incoming data as fast as possible, rather than wait for data to arrive on a few input links while others are idling.

Table 4.4 shows the resource utilisation of this implementation. The division of the HT into common memory structures such as individual arrays, and then individual daisy chained columns mean that the logic is much more localised than in the previously described implementations. This enables more placement and routing opportunities within the FPGA. For this reason, the daisy chain implementation is the most compact, but also the most scalable of the three options discussed.

4.3 Hough Transform Preprocessor

The HTP processes the 48-bit DTC stubs as they enter the TFP, both unpacking the data into a 64-bit extended format to reduce processing load on the HT, and assigning the stubs to geometric sub-sectors. This firmware is separate from the main HT. The HTP firmware consists of a preprocessing block, which calculates the correct sub-sector for each stub based on its global coordinate position and bend, followed by a layered routing block. The stubs associated with each sub-sector are routed to dedicated outputs, such that data from each sub-sector can be processed by an independent HT array.

Table 4.4 Resource utilisation of one column and of one entire HT array in the daisy-chained Hough transform, as implemented in the V7-690 [3]. The usage as a percentage of the available resources in the device are also shown. In total, the TFP needs 36 HT arrays. The total utilisation is given both with and without the MP7 infrastructure (infra.) firmware. A description of the resource units is in Sect. 1.4. As the daisy chain implementation relies on the HT preprocessor to pre-calculate the initial ϕ_T values, and the q/p_T range compatible with the stub bend, the additional DSP resource usage of the HTP must be taken into account when comparing this implementation with the systolic array

Firmware	LUT	DSP	FF	BRAM
1 column	118	2	204	1
1 HT array	6014	64	6718	33
1 HT mux	4431	0	4368	0
1 load balancing block	9473	0	24,290	0
18 HT arrays	108,000	1152	121,000	594
Load balancing block & MUX	14,000	0	29,000	0
MP7 infra.	96,000	0	890,000	330
Total	218,000	1152	239,000	924
Total (no infra.)	122,000	1152	150,000	594
Available in V7-690	433,000	3600	866,000	1470
Fraction of V7-690 [%]				
18 HT arrays	25	32	14	40
MP7 infra.	22	0	10	22
Total	60	32	27	62
Total (no infra.)	28	32	17	40

4.3.1 HTP Mathematics Block

The HTP firmware sorts all stubs that enter the TFP into 36 sub-sectors that loosely correspond to regions in η and ϕ (the exact mathematics is described later in this section). As shown in Fig. 4.11, these sub-sectors are formed from two divisions in the $r - \phi$ plane per nonant (or octant) and eighteen divisions in the $r - z$ plane. The division of the data into sub-sectors simplifies the task of the downstream logic, so that track finding can be carried out independently and in parallel within each of the sub-sectors. The use of relatively narrow sub-sectors in η has the added advantage that it ensures that any track found by the HT stage must be approximately consistent with a straight line in the $r - z$ plane, despite the fact that the HT itself only does track finding in the $r - \phi$ plane.

Each sub-sector contains stubs from different ranges in ϕ_T and z_S, where ϕ_T (z_S) is defined as the ϕ (z) coordinate of a track trajectory relative to the point where it crosses a cylinder of radius T (S) centred on the beam line. ϕ_T is the same parameter

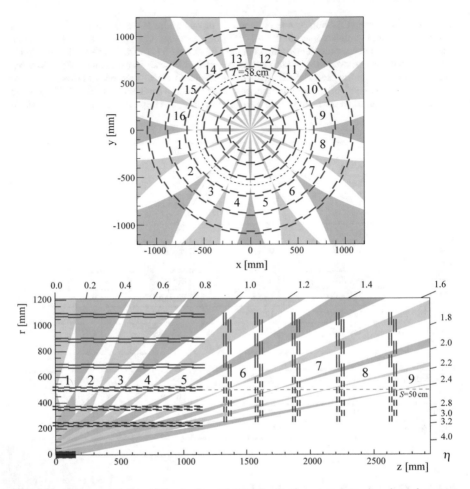

Fig. 4.11 The segmentation of the tracker volume into ϕ sub-sectors (upper) and η sub-sectors (lower), as used within the firmware implementation of the Hough transform preprocessor [5]. The numbered areas in white represent the regions that are associated with only one sector, whereas the coloured areas (where there is no difference in meaning between green or blue) represent the overlap region between neighbouring sectors where stubs may need to be assigned to both sectors. The two cylinders mentioned in the text of radius $T = 58$ cm and $S = 50$ cm, are indicated by dashes in the top and bottom figures, respectively

that was defined in Sect. 4.1 for use in the HT. For convenience, the same value of T is used for both the HT and the sub-sector assignment, but this is not a requirement. One can calculate z_S from a sector boundary in η with the formula

$$z_S = S \cot \left(2 \arctan e^{-\eta}\right). \tag{4.7}$$

The HTP must assign each stub to a sub-sector based on whether the stub could have been produced by a charged particle with a trajectory within the ϕ_T or z_S range of that sub-sector while originating from the beam line. If the stub is consistent with more than one sub-sector, then the HTP duplicates it. This can occur because of the curvature of tracks within the magnetic field (constrained by $p_T^{\min} = 2\,\mathrm{or}\,3\,\mathrm{GeV}$) or because of the length of the luminous region along the beam axis (where a configurable parameter w, chosen to be $15\,\mathrm{cm}$, defines the half-width of the beam spot along z).

A stub is compatible in the $r - z$ plane with a sub-sector that covers the range $z_S^{\min} < z_S < z_S^{\max}$ if

$$\frac{r \cdot z_S^{\min}}{S} - w \cdot \left|\frac{r}{S} - 1\right| < z < \frac{r \cdot z_S^{\max}}{S} + w \cdot \left|\frac{r}{S} - 1\right|. \tag{4.8}$$

To further improve the performance of the HT without using significant additional FPGA resources, each of these η sub-sectors can be further divided by an additional factor of two in the $r - z$ plane. This division is positioned at the mid-point between the sub-sector's boundaries ($z_S^{\min}$, $z_S^{\max}$). Whenever a stub is assigned to a sub-sector, the HTP checks the consistency of the stub with each of these sub-sector halves, allowing for some overlap, and stores this information as two bits within the stub data, for subsequent use by the HT.

The corresponding equation for the compatibility in the $r - \phi$ plane of the stub with a sub-sector is

$$|\Delta\phi| < \frac{\pi}{N_\phi} + \phi_{\mathrm{res}}, \tag{4.9}$$

where $\Delta\phi$ is the difference in azimuthal angle between the stub and the centre of the sub-sector and N_ϕ (which equals 18 for nonants or 16 of octants) is the total number of ϕ sub-sectors (which must be a multiple of the number of tracker sector divisions). The azimuthal angle of the centre of sub-sector k is $\phi_k = (2k - 1)\pi/N_\phi$, where $1 \leq k \leq N_\phi$. The parameter ϕ_{res} accounts for the range of track curvature in ϕ allowed by the threshold $p_T^{\min}$, and is equal to

$$\phi_{\mathrm{res}} = \left(\frac{B\,c\,|r_T|}{2}\right) \frac{q}{p_T^{\min}}. \tag{4.10}$$

With $N_\phi = 16$ no individual stub can be compatible with more than two neighbouring ϕ sub-sectors, providing that $p_T^{\min}$ is not reduced below $2\,\mathrm{GeV}$.

However, the stub can also be tested against a second condition in the $r - \phi$ plane, to reduce the number of stubs that need to be duplicated. This test exploits the stub bend measurement b, measured in units of the strip pitch, which is provided by the p_T-modules. The bend further constrains the allowed q/p_T range to that given by Eq. 4.5. As there are only eight possible values of (p/s), this quantity is retrieved from a look-up table in firmware. This equation assumes that the resolution in the bend, when measured in units of the sensor pitch, is the same everywhere in the

Table 4.5 η sub-sector boundary specification, as implemented in the demonstrator Hough transform preprocessor. The corresponding z_{50} value is also given, to the nearest cm

η region boundaries	2.4	2.16	1.95	1.7	1.43	1.16	0.89	0.61	0.31	0.0
z_{50} [cm]	273	214	172	132	98	72	51	32	16	0

tracker. Simulations confirm this assumption to be valid and indicate an approximate value of $1/\sqrt{6}$ for the resolution. The true bend is assumed to lie within k_b of the measured value, where k_b is a configurable parameter whose value is chosen to be 1.25 (approximately three standard deviations from the mean).

This constraint on q/p_T leads to the condition:

$$|\Delta\phi + b\rho\frac{r_T}{r}| < \frac{\pi}{N_\phi} + \phi'_{\text{res}} \tag{4.11}$$

where ϕ'_{res}, which allows for the finite resolution of the stub bend, is given by

$$\phi'_{\text{res}} = \min\left(k_b\,\rho\left|\frac{r_T}{r}\right|, \frac{\pi}{N_\phi}\right). \tag{4.12}$$

In the default HTP, S is chosen to be 50 cm, since this reduces the fraction of stubs that are consistent with more than one sub-sector. The ranges in ϕ_T or z_S covered by neighbouring sub-sectors are contiguous and do not overlap. In the $r - \phi$ plane, the sub-sectors are all equally sized, whereas in the $r - z$ plane their size varies so as to keep the number of stubs approximately equal in each sub-sector. The boundary definitions of the sub-sectors in the $r - z$ plane are given in Table 4.5. Each stub is assigned to an average of 1.8 sub-sectors.

4.3.1.1 HTP Routing Block

The routing block is implemented as a three-stage, highly pipelined mesh, as depicted in Fig. 4.12. It can route stubs from up to 72 inputs, one per DTC (with up to 36 DTCs assumed in each of the two detector octants from which the HTP receives data), to any of 36 outputs, where each output corresponds to a sub-sector. The first layer organises stubs into six groups of three sub-sectors in η, which in turn are each arranged according to their final η sub-sector in the second layer. The third layer routes the stubs by ϕ sub-sector. Each arbitration block in this router is highly configurable, and can easily be adapted for alternative sub-sector boundaries.

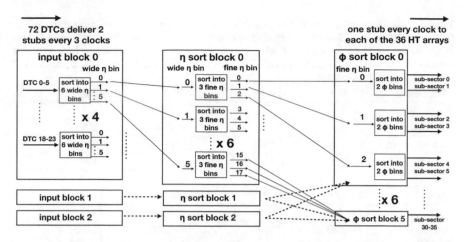

Fig. 4.12 Block diagram of the Hough transform preprocessor router. Each solid arrow corresponds to a connection that transports one stub per clock. Each dotted arrow corresponds to a bundle of solid arrows. This is based on a model where 72 DTCs deliver 48 stubs (48-bit wide) per 240 MHz clock to each TFP. Rough sorting in η is followed by fine sorting in η and finally by fine sorting in ϕ. Three input blocks each take stubs from 24 DTCs, and sort them into six wide bins in η. Each input block is connected to one and only one η sort block, where a fine sort in η takes place, resulting in 18 η bins. There are three η sort blocks in total. All η sort blocks are connected all six ϕ blocks. Each ϕ sort block sorts the stubs into two ϕ bins. The result is thirty six bins in total, six from each ϕ block. Each of these corresponds to one sub-sector, and stubs in each bin are therefore forwarded to their individual (and corresponding) HT array

4.3.1.2 HTP Resource Usage and Latency

The HTP mathematics block and router for one TFP can be implemented within a single V7-690 FPGA. The FPGA resource usage is shown in Table 4.6. Running at 240 MHz, the latency (defined as the time difference between first stub received and first stub transmitted) of the preprocessing and routing blocks is 58 and 193 ns, respectively. A version of the HTP router has been developed to run at 480 MHz. In this version, additional registers were required to successfully meet timing constraints, leading to an overall latency reduction of 60 ns. The HTP mathematics block is able to run at up to 500 MHz in the V7-690.

4.4 Hough Transform Results

While the standalone HT described here does not constitute a full track finder, due to the large number of fake track candidates, and the lack of a three dimensional track parameter fit, it is important to understand the potential of the HT for massively reducing the load on the downstream fitting, and how this potential compares to alternative approaches.

Table 4.6 Resource utilisation of the Hough transform preprocessor as implemented in the V7-690 FPGA. Resource utilisation of each HTP mathematics (math.) block (with 48 needed for the demonstrator TFP) and the routing block is given. The usage as a percentage of the available resources in the device are also shown. The total utilisation is given both with and without the MP7 infrastructure (infra.) firmware

Firmware	LUT	DSP	FF	BRAM
1 math. block	1942	22	2416	1
1 routing block	27,700	0	89,531	174
48 math. blocks	93,000	1056	116,000	48
1 routing block	28,000	0	90,000	174
MP7 infra.	96,000	0	89,000	330
Total	232,000	1056	295,000	552
Total (no infra.)	136,000	1056	206,000	222
Available in V7-690	433,000	3600	866,000	1470
Fraction of V7-690 [%]				
48 math. blocks	21	29	14	3
1 routing block	7	0	10	12
MP7 infra.	22	0	10	22
Total	50	29	34	37
Total (no infra.)	28	29	24	15

As the HT is not the final track fit, a definition of HT tracking efficiency is used that does not exactly match the definition used for the full chain tracking efficiency, as described and presented in Chap. 6. The *HT tracking efficiency* is the fraction of charged particles (within kinematic acceptance: $p_T > 3$ GeV, $|z_0| < 30$ cm, and transverse distance from beam line to particle vertex $L_{xy} < 1$ cm) that can be associated with a track candidate made up of a minimum of four stubs that were generated by this charged particle. In order to contribute to the efficiency denominator, the charged particle must have generated stubs in at least four different tracking layers. For the HT tracking efficiency, no requirement is set on the number of incorrect stubs that may be included with the track candidate, as these incorrect stubs are likely to be removed by the downstream stages. In simulated $t\bar{t}$ events with 200 pileup (flat geometry), the mean percentage of stubs on each track candidate that originate from the same genuine particle is 84%, and over 50% of the candidates have at least one unwanted (incorrect) stub.

Figure 4.13 shows the charged track multiplicity for a subset of tracks that are considered for efficiency metrics, as a function of $1/p_T$ and η. In this typical sample of $t\bar{t}$ and 200 pileup, the mean number of tracks per event that fit this criteria is 59. Of these, an average of 16 are from the $t\bar{t}$ vertex, the rest are from minimum bias pileup. These numbers can be compared to what is achieved by the track finder. Note that

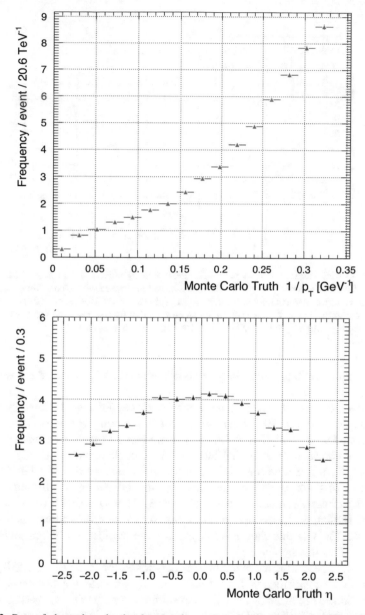

Fig. 4.13 Rate of charged tracks that leave at least one stub in a minimum of four tracker layers/disks. A standard sample of t̄t and 200 pileup is used. Tracks are included if they have a true p_T above 3 GeV, a true $|\eta| < 2.4$, a maximum transverse (longitudinal) impact parameter of 5 cm (30 cm). Frequency per event is shown as a function of $1/p_T$ (upper) and η (lower). In this typical sample of t̄t and 200 pileup, the mean number of tracks per event that fit this criteria is 59. Of these, an average of 16 are from the t̄t vertex, the rest are from minimum bias pileup

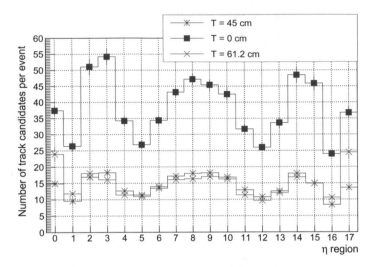

Fig. 4.14 Average number of track candidates generated by the HT for a sample of t$\bar{\text{t}}$ and 200 pileup, in the tilted geometry scenario. These results are generated with emulation of the default daisy chain Hough transform implementation. Results are shown with the radial offset of the Hough transform configured to 0, 45 and 61.2 cm. It can be seen that using a radial offset, as described in Sect. 4.1, significantly improves performance. The efficiency is preserved

although minimum bias multiplicity peaks at high η, the tracks from the t$\bar{\text{t}}$ vertex are most likely at low η.

Figure 4.14 shows the average number of track candidates generated by the HT for a sample of t$\bar{\text{t}}$ and 200 pileup, in the tilted geometry scenario. These results are generated with emulation of the daisy chain Hough transform implementation. Results are shown with the radial offset of the Hough transform configured to 0, 45 and 61.2 cm. It can be seen that using a radial offset, as described in Sect. 4.1, the number of track candidates generated by the HT is significantly reduced. For the tilted geometry scenario, a value of $T = 61.2$ cm was found to be optimum, given the limitations in the stub gradient due to firmware limitations (as described in Sect. 4.2.1). With this chosen value, the average number of track candidates per event is 270 for this sample.

Figure 4.15 shows the reduction in track candidates that can be obtained by utilising the stub bend information. In the tilted geometry scenario, the average number of track candidates produced by the HT per event for a sample of t$\bar{\text{t}}$ events at 200 pileup is 271 with the bend filter, and 808 without the bend filter. It should be noted that the bend filter can be disabled in the HT firmware with a single configurable parameter, if required. Disabling the bend filter does increase the rate of fake and duplicate track candidates that are reconstructed by the HT, however, even when including truncation effects, the average HT tracking efficiency for this sample improves from 97.1 to 97.9% when it is disabled. This is because of a small number of stubs with bend measurements far enough from their true p_T are not binned in the correct q/p_T

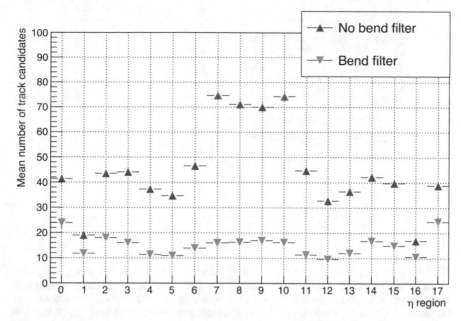

Fig. 4.15 Average number of track candidates generated by the HT for a sample of t̄t and 200 pileup, in the tilted geometry scenario. These results are generated with emulation of the baseline daisy chain Hough transform implementation. Results are shown with and without the HT bend filter option. It can be seen that using the bend filter, as described in Sect. 4.1 significantly improves performance

array column. The baseline configuration does utilise the bend filter, however, as it improves the purity of the track candidates, and reduces the load on the downstream processing during particularly busy events.

Figure 4.16 shows the degradation in tracking efficiency that results from choosing a too fine granularity of the array. In the finer granularity array presented, each cell has a width in ϕ_T of only 3.07 mrad. This is not wide enough to account for multiple scattering, bremsstrahlung, and other interactions that change the particle trajectory from a precisely straight line in the $r - \phi$ axis. In particular, this effect is most evident at low p_T, and becomes even more significant if tracking down to 2 GeV is required.

As shown in Fig. 4.17, a global requirement of stubs in four tracking layers for a track candidate to be selected by the HT increases the mean number of track candidates by a factor of about 4.25. This accompanies an increase in the HT tracking efficiency by a few percent in the regions around $0.5 < |\eta| < 1.5$. A large increase in the number of track candidates at high η means that the four layer requirement option produces a lower efficiency in this region, due to data truncation (no time to read out all track candidates). As a result, the default configuration is only to reduce the requirement where necessary, in the region, $0.89 < |\eta| < 1.16$.

The primary accomplishment of the HT is in the rate reduction and pileup removal that is accomplished (without significant efficiency loss for genuine tracks). For

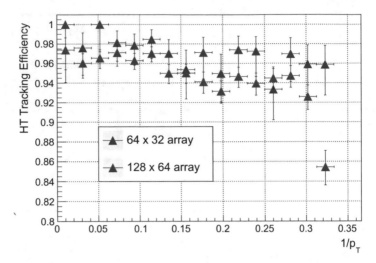

Fig. 4.16 Efficiency as a function of $1/p_T$ in the HT for a sample of $t\bar{t}$ and 200 pileup, in the tilted geometry scenario. These results are generated with emulation of the default daisy chain Hough transform implementation. Results are shown with the number of cells in the Hough transform array configured to 128 (64) in ϕ and 64 (32) in q/p_T per sub-sector for the blue (and red) plots. It can be seen that in particular tracking efficiency at low p_T is reduced when the chosen cell granularity is too fine

example, in the flat geometry for $t\bar{t}$ samples at 200 pileup, there are an average of approximately 16,000 stubs per event at input to the TFP. The HTP must duplicate stubs when sorting into sub-sectors, and so this becomes 25,600 stubs at input to the HT. After the HT, the total number of stubs on track candidates remaining is 2300. With the standard 240 MHz HT, this rate reduction by a factor of 7 (11 if including the HTP duplication) is accomplished in about 1 μs. The downstream track fitting is therefore able to avoid the huge combinatorial problem that it would be exposed to if the stubs were not pre-filtered, and pre-sorted. This allows for a much more thorough track fitting approach downstream, such as the Kalman filter.

4.4.1 Optimisations and Improvements

The radial offset parameter, T, is fixed at 58 cm in the demonstrator for the entire tracker solid angle. However, this choice is not optimal at high pseudorapidities, where particles will not traverse the full radial extent of the tracker. At these high $|\eta|$ values, a smaller value of T would be preferred, as it would spread the gradients of the stub lines in Hough space equally between positive and negative values. Reducing the value of T down to 47 cm in the highest $|\eta|$ sectors reduces the rate of track-candidates produced by the HT in these sectors by a factor of two. This change

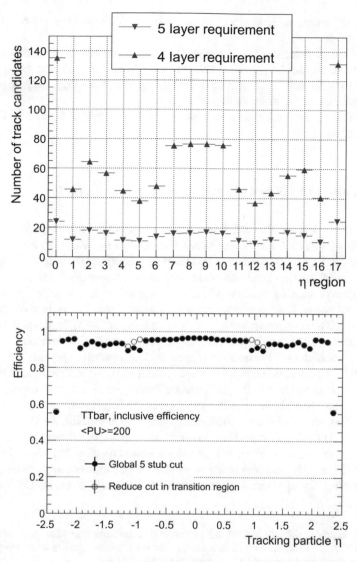

Fig. 4.17 Average number of track candidates generated by the HT for a sample of t̄t and 200 pileup, in the tilted geometry scenario (upper), where a requirement is set for either 5 or 4 layers. It can be seen that a four layer requirement increases the mean number of track candidates by a factor of 4.25. Tracking efficiency for a sample of t̄t and 200 pileup (lower), with the requirement for 5 layers relaxed only in the region $0.89 < |\eta| < 1.16$. Due to the gap in coverage in the region, the tracking efficiency improves

would be trivial to implement in firmware, without additional latency or significant resource usage.

The effect of using a grid of hexagonal or diamond cells in the HT instead of a conventional, rectilinear grid has also been studied in software, and indicates that a reduction in the rate of track candidates by approximately 20% may be possible, whilst maintaining the track finding efficiency.

More HT results, including a discussion of track finding down to 2 GeV, can be found in Chap. 6.

4.5 Scaling to Ultrascale and Ultrascale+ FPGAs

As the Ultrascale and Ultrascale+ generation of FPGAs, such as the Xilinx Virtex Ultrascale 9+ (VU-9P) and Xilinx Virtex Ultrascale 11+ (VU-11P) are compatible with up to 25 Gbps transceivers, one must consider how the HT designed would scale to a system with such an input bandwidth. In addition, faster clock frequencies equate to faster data processing which would allow fewer resources per node (or fewer time multiplexed nodes) to be required, reducing the number of FPGAs, boards and cost of the system. If the HT firmware was capable of running at 480 MHz, it would be able to take in data from 16 Gb/s links without major changes to the implementation. To this end, a version of the HT has been developed, with an optimised pipeline for clocking at 480 MHz. Two additional pipeline steps (which lead to a 30% increased Flip Flop utilisation per array) means that the pipeline latency becomes 92 ns (comparable to 175 ns for the 240 MHz design). Table 4.7 gives the full resource usage numbers of this 480 MHz variant of the HT array, as implemented in the Xilinx Kintex Ultrascale 115 (KU-115) FPGA. It should be noted that although one HT array is able to run at 480 MHz in this chip, clock skew issues associated with clocking the DSPs that are used to facilitate adjacent bins in firmware that are not adjacent q/p_T prevent 32 HT arrays meeting timing above 416 MHz currently. It is likely that these DSPs could be

Table 4.7 Resource utilisation of the daisy chain HT, as optimised for 480 MHz running in the KU-115 FPGA. This implementation does not include any infrastructure firmware, except for two transceivers/receivers running at 16 Gb/s connected to each HT array

Firmware	LUT	DSP	FF	BRAM
One HT array	5750	64	7500	32
32 HT arrays plus MUX	184,000	2042	240,000	1036
Available in KU-115	633,000	5520	1,266,000	2160
Fraction of KU-115 [%]				
32 HT arrays plus MUX	29	37	19	48

removed by including additional constant terms in the calculations. It has also been observed that timings are improved by about 10% when building the implementation for a Virtex Ultrascale chip, which is expected as a consequence of Xilinx making improvements in their chip design.

Figure 4.18 depicts how a set of HT arrays could be implemented with 25 Gbps input links, where the HT is running at a frequency of 480 MHz or less. An internal time-multiplexing within the HT FPGA would enable each array to see data coming in at a lower clock frequency, and have the required time to process it. With a factor of two internal time-multiplexing, the general formula for the number of stubs that can be processed per array would be

$$N_{stubs} = \frac{168 \times 2f}{480}, \tag{4.13}$$

where f is the frequency of the HT array. Following this simple scaling, the required number of HTs per TFP would be given by

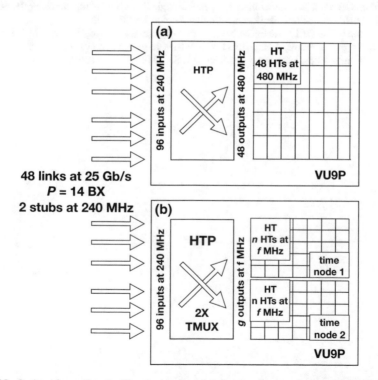

Fig. 4.18 Options for scaling the Hough transform implementation to a design with 25 Gb/s input links. **a** Depicts an option where the HT is able to run at 480 MHz, and the scaling is therefore very simple. **b** Shows a scenario where the HT runs at a frequency less than 480 MHz, and an internal factor of two time multiplexing is required to process the incoming data

$$N_{\text{HTs}} = 2 \times \text{ceil} \left(\frac{480 \times 48}{f} \right).$$ (4.14)

This is depicted by (b) in Fig. 4.18.

References

1. Hough PVC (1962) Method and means for recognizing complex patterns, US Patent 3,069,654, Dec 1962. http://citeseerx.ist.psu.edu/viewdoc/download?doi=10.1.1.85.8417& rep=rep1&type=pdf
2. Duda RO (1972) Use of the Hough transformation to detect lines and curves in pictures. Commun Assoc Comput Mach 15(1):11–15. https://www.cse.unr.edu/~bebis/CS474/Handouts/ HoughTransformPaper.pdf
3. Xilinx Inc (2017) 7 series FPGAs data sheet: overview, product specification, DS180 (v2.5). https://www.xilinx.com/support/documentation/data_sheets/ds180_7Series_Overview.pdf
4. Schuh T (2017) Entwicklung des CMS-spurtriggers für den Hochluminositätsbetrieb des Large Hadron Colliders. Karlsruher Institut für Technologie Ph.D. thesis, https://doi.org/10.5445/IR/ 1000079109. https://publikationen.bibliothek.kit.edu/1000079109
5. Aggleton R et al (2017) An FPGA based track finder for the L1 trigger of the CMS experiment at the High Luminosity LHC. JINST 12:P12019. https://doi.org/10.1088/1748-0221/12/12/ P12019. http://iopscience.iop.org/article/10.1088/1748-0221/12/12/P12019/pdf

Chapter 5
The Kalman Filter

5.1 The Kalman Filter Algorithm

The KF [1, 2], uses a technique known as linear quadratic estimation to produce
estimates of unknown variables from a series of uncertain measurements. Bayesian
inference is used to estimate a joint probability distribution whenever a new measurement is taken. It is widely used in scientific and technological fields. Examples
include aircraft and spacecraft navigation [3, 4], trajectory optimisation, signal processing [5], and econometrics [6].

5.1.1 The Generic Kalman Filter

The KF is a recursive estimator, allowing new measurements to be processed as they
arrive. This section will give the generic formulation and description of the Kalman
filter. The section that follows will then give a derivation for the application to track
fitting. Figure 5.1 is an illustration that shows the flow of information in the Kalman
filter. The variables and matrices shown are explained in the following text.

The derivation for a generic Kalman filter begins by defining a state vector x which
contains some information about the system that is being studied. This state is an
unknown, and the role of the Kalman filter is to find the best estimate of the elements of
x, the state parameters. The Kalman filter assumes that these parameters are Gaussian
distributed (but they do not have to be). The parameters in x are also related by a
symmetric covariance matrix, $\mathbf{P}$ which represents the degree of correlation between
each and every parameter. It is possible to define the best estimate of x at time t, $\hat{x}_t$,
with covariance matrix $\mathbf{P}_t$. It should be noted that t does not necessarily represent
time, but any stepping parameter. Given a set of modelling assumptions about the
behaviour of x as steps are taken in t, it is possible to write down a prediction matrix
$\mathbf{F}_t$ such that

$$\hat{x}_{t+1|t} = \mathbf{F}_t \hat{x}_{t|t}. \tag{5.1}$$

© Springer Nature Switzerland AG 2019
T. O. James, *A Hardware Track-Trigger for CMS*, Springer Theses,
https://doi.org/10.1007/978-3-030-31934-2_5

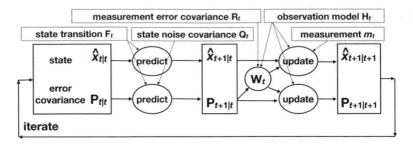

Fig. 5.1 Illustration of the flow of information in the Kalman filter

The covariance of any distribution behaves the same way when the underlying distribution is transformed. In this case, the covariance matrix evolves such that

$$\mathbf{P}_{t+1|t} = \mathbf{F}_t \mathbf{P}_{t|t} \mathbf{F}_t^{\mathsf{T}}. \tag{5.2}$$

Unknown influences on the state, such as noise, are assumed to be Gaussian, with a covariance $\mathbf{Q}_t$. These do not affect the state estimate, but they do affect the state covariance. This term can therefore be added to the equation for $\mathbf{P}_{t+1}$. In addition, to account for known external forces that affect the state (but not the covariance), a term can be added to the equation for $\hat{x}_{t+1}$. This term consists of a control vector, u_t, that transforms with the matrix $\mathbf{B}_t$ between steps. The equations for the prediction step are then as follows:

$$\hat{x}_{t+1|t} = \mathbf{F}_t \hat{x}_{t|t} + \mathbf{B}_t u_t,$$
$$\mathbf{P}_{t+1|t} = \mathbf{F}_t \mathbf{P}_{t|t} \mathbf{F}_t^{\mathsf{T}} + \mathbf{Q}_t. \tag{5.3}$$

These equations give a new best estimate $\hat{x}_{t+1}$, given a previous best estimate $\hat{x}_t$ and a correction for known external influences $\mathbf{B}_t u_t$. They also give a new uncertainty matrix $\mathbf{P}_{t+1|t}$, given a previous uncertainty matrix $\mathbf{P}_{t|t}$, and some additional (Gaussian) unknown uncertainty $\mathbf{Q}_t$. The next step is to improve this estimate by including measurement data: the vector m_t, with an associated uncertainty, $\mathbf{R}_t$. This can be done by defining a matrix $\mathbf{H}_t$, that models the expected measurements at t such that the expected measurement $\hat{m}_t = \mathbf{H}_t \hat{x}_{t+1}$. The two Gaussian distributions, around m_t and $\hat{m}_t$ both contain information that is useful to improve the estimate of x. They can be summarised by the following, where μ and Σ are the associated mean and covariance of the distributions:

$$(\mu_0, \ \Sigma_0) = \left(\mathbf{H}_t \hat{x}_{t+1|t}, \ \mathbf{H}_t \mathbf{P}_{t+1|t} \mathbf{H}_t^{\mathsf{T}} \right),$$
$$(\mu_1, \ \Sigma_1) = (m_t, \ \mathbf{R}_t). \tag{5.4}$$

Standard Gaussian multiplication can be used to construct a new distribution from the overlap of these two. This operation can be written as

$$\mu' = \mu_0 + \mathbf{K}\,(\mu_1 - \mu_0),$$
$$\Sigma' = \Sigma_0\,(1 - \mathbf{K}) \tag{5.5}$$
$$\text{where } \mathbf{K} = \Sigma_0\,(\Sigma_0 + \Sigma_1)^{-1}.$$

By applying Eqs. 5.5–5.4, one derives the complete equations for the update step of the Kalman filter.

$$\hat{x}_{t+1|t+1} = \hat{x}_{t+1|t} + \mathbf{W}_t v_t,$$
$$\mathbf{P}_{t+1|t+1} = \mathbf{P}_{t+1|t} - \mathbf{W}_t \mathbf{H}_t \mathbf{P}_{t+1|t}, \tag{5.6}$$

where the matrix known as the Kalman Gain

$$\mathbf{W}_t = \mathbf{P}_{t+1|t}\mathbf{H}_t^{\mathrm{T}}\left(\mathbf{H}_t \mathbf{P}_{t+1|t}\mathbf{H}_t^{\mathrm{T}} + \mathbf{R}_t\right)^{-1}, \tag{5.7}$$

the measurement residual

$$v_t = m_t - \mathbf{H}_t \hat{x}_{t+1|t}, \tag{5.8}$$

and $\hat{x}_{t+1|t+1}$ is the new best estimate of the state, with covariance $\mathbf{P}_{t+1|t+1}$. This process of prediction followed by updating with new measurements can be repeated as many times as required.

5.1.2 The Kalman Track Fitter

The choice to use a Kalman filter for track fitting in the demonstrator was motivated by the features of the track candidates presented by the Hough transform. These candidates consist of a series of stubs that loosely (but not precisely) correspond to a realistic track in CMS. Many track candidates contain stubs that did not originate from the simulated particle that generated the majority of stubs within the candidate. In typical $t\bar{t}$ events with 200 pileup, over half of otherwise genuine track candidates contain at least one such stub. In addition to this, many track candidates correspond to combinations of stubs that do not correspond to a real particle trajectory, but are binned in the same HT cell due to the limited resolution of the HT, and the high stub occupancy within the tracker. For this same sample, in the tilted (flat) barrel geometry, 25% (44%) of the candidates fall into this category. With this in mind, the Kalman filter is potentially a good choice for the next step in the track finding chain. The formalism described in Sect. 5.1.1 can be applied to the track fitting process, where the Kalman state can be defined as the current best estimate of the fitted track parameters, and the measurements to be added are the individual stubs contained within each track candidate. One can therefore apply the Kalman formalism to a track parameter estimate (or prior), given by the HT, and add successive stubs in order of increasing radius in turn, until all stubs are added, and a final track parameter estimate is obtained. A discussion of track reconstruction with the Kalman filter can be found in [1, 7–10], and it has been widely used in particle physics, including in the CMS offline reconstruction up until now.

A desirable feature of this particular Kalman filter implementation would be both to remove unwanted stubs from track candidates prior to the final fit (thus improving the fitted track parameters), and discard track candidates that do not correspond to a genuine track. Both of these tasks must be accomplished without significant efficiency losses caused by the discarding of genuine tracks or desirable stubs. A firmware architecture was therefore designed that would allow this.

Figure 5.2 illustrates the process of adding (or rejecting) stubs in the track fit. The filter begins with an estimate of the track parameters and their uncertainties, also referred to as the *state*, $\hat{x}_{t|t}$. Stubs are used, iteratively, to update the state following the Kalman formalism, decreasing the uncertainty in the state with each measurement. A weighting derived from the relative uncertainties in the state and measurement, the Kalman gain $\mathbf{W}_t$, controls the adjustment of the track parameters.

For the TFP demonstrator, an implementation was chosen that fits the following four parameters:

$$x = (1/2R, \ \phi_0, \ \cot\theta, \ z_0) \,, \tag{5.9}$$

where R is the track radius of curvature and related to q/p_T according to Eq. 2.2, ϕ_0 is the azimuthal angle of the track in the transverse plane at the beam line, θ is the polar angle and z_0 is the longitudinal impact parameter.

An aspect of the KF that makes it suitable for FPGA implementation is the relatively small size of the matrices which must be manipulated (4×4 matrices for a 4 parameter fit). The size of these matrices is independent of the number of measurements, minimising logic usage. The iterative nature of the KF increases the required

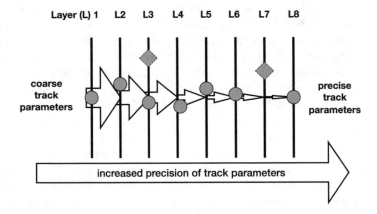

Fig. 5.2 An illustration of the process of iteratively adding stubs to the Kalman fit in order of increasing radius. As more stubs are added, the state estimate of the track parameters becomes more precise. The grey circles represent stubs that are included in the fit, and the grey diamonds with dotted borders represent stubs that are rejected as they are too far from the current track estimate

complexity of the FPGA firmware, however, alternative track fitting procedures are also iterative in nature, but in a different way. The KF approaches the final values of the state parameters by adding successive measurement points, which force the parameters to change. In contrast, linear regression-based track fit methods solve a set of equations by successive approximations to the fit values. This fitting procedure means that the size of the matrices (and subsequently the resources needed to manipulate them) linearly increase with the number of hits (stubs) that are being used to fit the track. This is not favourable for an FPGA solution, where the algorithms must fit within resource usage constraints, and the size of the matrices is ideally pre-determined.

Assuming that tracks originate at $r = 0$, a simplified form of the track equations can be expressed in linear terms of the stub radius r,

$$\phi = \phi_0 - \frac{r}{2R},$$
$$z = z_0 + r \cot \theta. \tag{5.10}$$

If $m(\phi, z)$ is the measurement, the update matrix $\mathbf{H}_t$ (found by taking the partial derivative of the track equations with respect to the elements of x), can be expressed as

$$\mathbf{H}_t = \frac{\partial m}{\partial x} = \begin{bmatrix} -r & 1 & 0 & 0 \\ 0 & 0 & r & 1 \end{bmatrix} \tag{5.11}$$

A further simplification is possible. As the true track parameters are not expected to change between subsequent measurements, the forecast matrix $\mathbf{F}_t$ is simply the identity matrix $\mathbf{I}_4$. As the current FPGA implementation of the KF does not take into account the effects of multiple scattering, or other white noise-like effects, the matrix $\mathbf{Q}_t = 0$. This means that $\mathbf{P}_{t+1|t} = \mathbf{P}_{t|t}$ and $\hat{x}_{t+1|t} = \hat{x}_{t|t}$.

$\mathbf{P}_{t|t}$ can be written as a symmetric 4×4 matrix, and is related to $\hat{x}$ in the following way:

$$x_t \, \mathbf{P}_{t|t} = (1/2R, \ \phi_0, \ \cot \theta, \ z_0) \begin{pmatrix} \sigma_a^2 & \sigma_{ab} & 0 & 0 \\ \sigma_{ab} & \sigma_b^2 & 0 & 0 \\ 0 & 0 & \sigma_c^2 & \sigma_{cd} \\ 0 & 0 & \sigma_{cd} & \sigma_d^2 \end{pmatrix}. \tag{5.12}$$

Eight elements are always zero, as the assumed track equations (and therefore update matrix $\mathbf{H}_t$) determine that $(1/2R, \ \phi_0)$ and $(\cot \theta, \ z)$ can be treated as independent linear fits. The initial (seed) values for σ_a, σ_b, σ_c, and σ_d are determined from the uncertainties in the Hough transform step. For example, σ_a^{seed} and σ_b^{seed} are given by the uncertainty of the q/p_T and ϕ_0 estimates from the HT. These can be determined by calculating $1/\sqrt{12}$ of the corresponding HT cell widths. In addition, σ_c^{seed} is determined by $1/\sqrt{12}$ of the width of the η sub-sector that the track candidate was found in. On the other hand, σ_d^{seed} is the RMS of the beam spot width (5 cm).

Uncertainties σ_{ab} and σ_{cd} are initialised to zero, but will evolve over subsequent iterations.

Uncertainties in the stub measurements are written in the form of the matrix

$$\mathbf{R}_t = \begin{bmatrix} \sigma_\phi^2 & 0 \\ 0 & \sigma_z^2 \end{bmatrix} \tag{5.13}$$

where in the barrel

$$\sigma_\phi^2 = \left(\frac{1}{\sqrt{12}} \frac{p}{r} \right)^2, $$

$$\tag{5.14}$$

$$\sigma_z^2 = \left(\frac{1.5625\, l}{\sqrt{12}} \right)^2, $$

and in the endcaps

$$\sigma_\phi^2 = \left(\frac{1}{\sqrt{12}} \frac{\rho}{r} \right)^2 + \left(\frac{1.05}{2R} \right)^2, $$

$$\tag{5.15}$$

$$\sigma_z^2 = \left(\frac{1.5625\, l}{\sqrt{12}} \right)^2 0.9\, (\cot \theta)^2, $$

where here p is the strip pitch, and l is the strip length: 5 cm in the 2S modules; and 1.5 mm in the PS modules. The values of $1.05/2R$ used here are based upon the q/p_T HT bin of the track candidate, implemented with a look up table.

The full set of KF calculations to solve at each step in t (which equates to the incorporating of another stub in order of increasing radius) therefore simplify down to:

$$\mathbf{P}_{t+1|t+1}(r) = \mathbf{P}_{t|t} - \mathbf{W}_t \mathbf{S}_t \mathbf{W}_t^T,$$
$$\hat{x}_{t+1|t+1}(r) = \hat{x}_{t|t} + \mathbf{W}_t v_t, \tag{5.16}$$

where

$$\mathbf{S}_t(r) = \mathbf{H}_t \mathbf{P}_{t|t} \mathbf{H}_t^T + \mathbf{R}_t,$$
$$\mathbf{W}_t(r) = \mathbf{P}_{t|t} \mathbf{H}_t^T \mathbf{S}_t^{-1}. \tag{5.17}$$

It is evident from these formulae that $\mathbf{S}_t$ (a 2×2 matrix, as determined by the size of the measurement vector m_t, regardless of the number of iterations) is the only matrix to be inverted. As this is a 2×2 matrix, all elements in the inverse have the same denominator, so only one division is required. This is a major advantage over linear fits, where larger matrices must be inverted. Section 5.1.3 explains how it is possible to accomplish the above calculations within FPGA firmware.

The KF firmware implementation can be separated into two parts:

- A state updater block to carry out the matrix operations described by the Kalman formalism. This logic updates the state, including the track parameters, the covariance matrix, and the χ^2, along with additional parameters to be used for selection or filtering;
- A flow control firmware which must gather stub and state information to present it to the state updater, select and store updated states, and execute iteration cycles.

5.1.3 The Kalman State Updater

The Kalman state updater is the firmware block that executes the matrix mathematics. The required matrix mathematics is implemented in a fully pipelined manner (with the help of a higher level synthesis tool MaxJ [11, 12]). Other useful parameters such as the χ^2 of the track parameters, and the number of missing (no stub) layers are also calculated and stored. The KF state update is implemented in fixed-point arithmetic, not floating point calculations. This is required to keep resource usage reasonable such that the algorithm could fit in modern FPGA devices, and to keep the latency within the L1 trigger requirements. It was therefore required to study and tune the fixed point bit sizes and ranges in order to ensure the required track fitting precision was achieved. This was done in C++ simulation, and further alterations were informed by comparisons between firmware output, and floating point simulations.

The Xilinx Virtex 7 series of FPGAs offer several (fully pipelined) built-in options for multiplying high-bit size fixed-point numbers with DSPs. One 18-bit and one 25-bit quantity can be multiplied in a single DSP (taking three clock cycles), or one 25-bit and one 35-bit quantity can be multiplied using two DSP units (taking four clock cycles). To minimise the DSP resource usage, the former option was used wherever possible, however, the covariance matrix update path required higher precision than could be achieved with the 18-bit by 25-bit option, if it was to closely match floating point simulations. Therefore, this path utilises the high precision (25-bit by 35-bit) multiplication variant. The diagonals of $\mathbf{P}_t$ are therefore stored with 25 bits, and the intermediate quantities such as the elements of the inverted matrix $\mathbf{S}_t^{-1} = \left(\mathbf{H}_t \mathbf{P}_t \mathbf{H}_t^{\mathrm{T}} + \mathbf{R}_t\right)^{-1}$ and $\mathbf{W}_t$ are encoded with 35 bits. To minimise latency, the matrix multiplications are implemented in separate, parallel DSP instances and results that must be accumulated are done so with a maximally parallelised balanced adder tree. This tree is shown in Fig. 5.3.

The High Level Synthesis (HLS) language and compiler tool MaxJ [11–13] was used to implement the state updater part of the firmware, and the design benefits from the built-in fixed-point support and pipeline scheduling provided by the tool. A higher level tool like MaxJ is well suited to highly mathematical operations such as the state updater, as the tool simplifies the task of ensuring correct pipelining and timing for these operations, allowing the user to focus on correctly implementing the mathematics. The division which is needed for the matrix inversion has been

Fig. 5.3 The balanced adder tree, used to implement the matrix multiplication in the Kalman filter

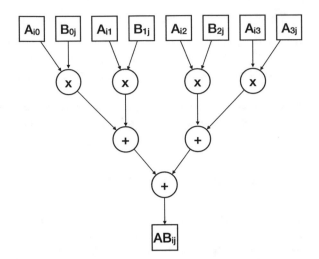

implemented in FPGA firmware in a custom, fast and lightweight way. A detailed description of this method can be found in [14].

The allowed range of many of the bit-wise parameters are determined directly from the tracker geometry and the boundary constraints imposed by the previous track finding stage, the HT. For example, the radius of curvature R must be larger than p_T^{min}/qB where by default, $p_T^{min} = 3$ GeV. In addition, in over 99% of cases, the track z_0 is expected to lie in the range $[-15\,\text{cm}, 15\,\text{cm}]$, as determined by the interaction region length. To allow a margin of error, the parameter ranges implemented in firmware are actually chosen to be twice that of these analytical results. This is necessary to allow intermediate steps in the calculation, where the parameters take non-physical values, or in the case of combinatorial fake tracks which may fit curves that point outside the expected ranges.

A minor firmware optimisation that has been implemented is to ignore elements in the covariance matrix that are always zero, and take advantage of its symmetric property. This means that only P_{01}, P_{23}, and the four diagonal elements, P_{ii} must be stored.

5.1.4   The Kalman Filter Flow Control

Figure 5.4 shows the fitting procedure for an example candidate. A seeded estimate for the state is obtained from the HT array index (q/p_T, ϕ_T) and η, ϕ sub-sector in which the track candidate was found. Starting with the seed state and its covariance matrix, stubs are used to update the state, ordered by increasing radius. To allow for detector inefficiencies or for the possibility that no compatible stub is found on a given layer, up to two non-consecutive layers may be skipped. In the case that a

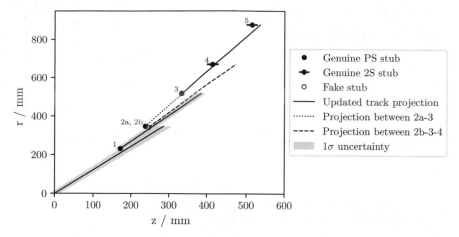

Fig. 5.4 An example of the Kalman filter fitting procedure for an HT candidate in the barrel, shown in the $r - z$ plane [14]. Genuine stubs are those associated with the same simulated charged particle, and fake stubs are those which are not. Line segments represent the fitted track trajectory at that point of the fit, updating with increasing radius, with the shaded area around the line showing one standard deviation of the track parameter estimate. Dashed track segments highlight the different result after fitting with stub 2a or 2b. The state that includes stub 2b is rejected after propagation to stub 4, due to failing a χ^2 cut in two consecutive layers

track candidate contains more than one stub on a given detector layer (which can occur by chance, or when detector elements overlap), each combination of stub and incoming state is propagated separately. This is why the implementation can be called a *combinatorial* Kalman filter. This eliminates any possibility of the incorrect stub affecting the fit of the genuine combination. The resulting states are ordered, giving preference first to states with fewer missing layers, and then with the smallest χ^2. Only the best state according to this measure is obtained from the filter, so no new duplicates are introduced.

One Kalman worker consists of a state updater, and the surrounding flow control firmware that manages the stub and state information to ensure that the state updater is provided with a steady flow of new and partially processed states; and the required stubs for each iteration. Figure 5.5 shows the connection of the logical elements within one KF worker, and their operation is described in the following text.

Stubs for a set of track candidates arrive in packets from the HT. Since the algorithm is iterative and an iteration takes many clock cycles, the stubs are stored in memory (FIFO 1) for later retrieval. The seed creator outputs the starting state in the required format.

The HT array index and sub-sector are used as a unique ID for each track candidate, providing a reference to the stubs stored in memory at the first step. Only one state can enter the state updater on each clock cycle, and there may be competition between partially worked states waiting in FIFO 3, and a new candidate arriving into the worker, whose seed state is waiting in FIFO 2. The state control block multiplexes

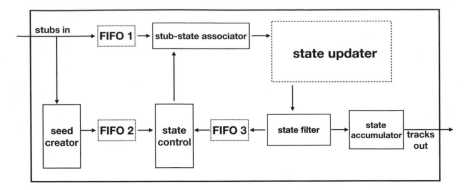

Fig. 5.5 Block diagram showing the connection of logical elements within a Kalman filter worker. The function of each of the firmware blocks is described in the text

the incoming states, giving preference to new candidates. The stub-state associator block uses the IDs stored with the state to retrieve associated stubs, in order to update the current state. It determines which iteration the current state is on and passes any stubs within the candidate assigned to the next layer, (or the next-to-next layer in the case of a skipped layer), one per clock cycle. Stubs from the next-to-next layer can only be forwarded to the state updater if the current state indicates that it has not skipped two layers already.

One iteration of the KF is executed in the state updater, using the new stub m_t and the current state $\hat{x}_{t|t}$. The new state estimate $\hat{x}_{t+1|t+1}$, $\mathbf{P}_{t+1|t+1}$, updated χ^2 value, and other status information are all calculated. At the output of the state updater, any states that fail a set of configurable cuts are discarded. The applied χ^2 cuts are 15, 100, and 320 for states with two, three, and four layers respectively. If the track p_T is below 2.7 GeV, the cuts for three and four layer states are relaxed to 120 and 1420 respectively. These cuts were tuned to maximise efficiency in simulation. In addition, a state with two layers or more is discarded if either of the following are true: an estimated $|z_0| > 15$ cm; or an estimated $p_T < p_T^{\min} - 0.1$ GeV, where $p_T^{\min}$ is the track finding p_T threshold (2 or 3 GeV), and 0.1 GeV is a configurable tolerance.

After these cuts are applied, the state filter is then able to select against the remaining states based on p_T, χ^2, z_0, and a minimum requirement on the number of stubs from PS modules (which ensures good z_0 resolution). At this point, states that have fitted parameters that lie outside of the η, ϕ sub-sector where the candidate was initially found are also discarded. The state filter preserves up to the best N output states (by χ^2), for a given state from the previous iteration. On the first iteration all possible states are kept (which is a maximum of 13 as there can only be 16 stubs in a track candidate); on subsequent iterations this is reduced to four. This helps minimise the total number of states circulating in the worker at once, whilst preserving the best state the majority of the time.

The surviving states are written into FIFO 3, to complete further iterations of the Kalman filter. A completed track is one where a state has finished four iterations of the KF, after which the state is no longer re-inserted into this FIFO. All states surviving the selection cuts are presented to the state accumulator, where the best state for each track candidate is stored until a time-out signal is propagated, and the fitted tracks are read out. In the accumulator, preference is first given to states with fewer missing layers, and then with the smallest χ^2. This block allows, if required, the readout of partially filtered states on receipt of the time-out. This may be useful when processing particularly dense jets, where many candidates (each containing many stubs) mean that all iterations may not be complete before the time-out.

5.2 Resource Usage and Latency

The resource usage of a single KF worker is summarised in Table 5.1. Thirty six KF workers can be implemented in parallel in the V7-690, running at 240 MHz. For the TFP demonstrator, 72 KF workers are implemented, one for each channel (two links transferring one stub per clock) from the pair of HT boards.

Table 5.1 Resource utilisation of the KF state updater, and one full KF worker, as implemented in the V7-690 [15]. The usage as a percentage of the available resources in the device are also shown. For each TFP, a total of 72 workers processing data from 36 HT arrays are used. As each DR block processes the track candidates associated with six sub-sectors, in total twelve duplicate removal blocks are needed per TFP. Four types of FPGA resources are given. A description of the FPGA and each type of logic resource can be found in Sect. 1.4. The total usage is given with and without the MP7 infrastructure (infra.) firmware

Firmware	LUT	DSP	FF	BRAM
1 state updater	4014	70	3094	6
1 Kalman worker	5520	71	4370	25
1 DR router	16,611	0	24,655	108
1 DR block	291	0	496	4
36 Kalman workers	199,000	2556	157,000	882
3 DR blocks	900	0	1500	12
MP7 infra.	96,000	0	89,000	330
Total	312,500	2556	272,000	1334
Total (no infra.)	216,500	2556	183,000	1004
Available in V7-690	433,000	3600	866,000	1470
Fraction of V7-690 [%]				
36 Kalman workers	46	71	18	60
MP7 infra.	22	0	10	22
Total	72	71	31	91
Total (no infra.)	50	71	21	68

Table 5.2 Percentage of tracks that fail to be fitted with four stubs after time-out in t$\bar{\text{t}}$ events with 200 pileup in the flat geometry. This configurable time-out, or accumulation period can range from 900 to 1800 ns. Several different values for this period were tested in hardware, and it was found that for this particular sample (over 2000 events), a period of 1550 ns was sufficient to allow all tracks to be fitted with four stubs

Accumulation period [ns]	900	1450	1500	1550
Fraction of tracks with < 4 stubs [%]	100	55	13	0

Each logical element in Fig. 5.5 is implemented with a fixed latency. The latency of a single KF iteration is dominated by the matrix operations involved in the state updater, which takes 55 clock cycles. With a 240 MHz clock frequency this is 230 ns. At each iteration, multiple stubs go into, and (after a 55 clock cycle delay) come out of, the state updater on subsequent clock cycles. Allowing independent propagation of multiple stubs on a layer slightly increases the total latency compared to just four passes of the single iteration latency. An accumulation period of 1550 ns before time-out is set, after which point all tracks, completed or uncompleted, for one event are output. Measurements shown in Table 5.2 suggest that fewer than 0.1% of tracks in top quark pair-production (t$\bar{\text{t}}$) events with pileup of 200 fail to be fully reconstructed within this accumulation period. Since the state keeps track of the current iteration (identical to the number of stubs on the state), quality cuts can be placed on the final tracks, if, for example, only completed KF tracks are required. All results also assume that any tracks that arrive 300 ns after the first track for that event are lost. In t$\bar{\text{t}}$ events with 200 pileup (flat geometry), this amounts to a loss of fewer than 0.1% of tracks. This way, the track-finder can guarantee the downstream processors that they will have to wait no longer than 300 ns after the first track arrives to have access to all the tracks from the event.

5.3   Potential for Improvements

The Kalman filter used in the demonstrator is in some respects a minimal version of the full Kalman filter algorithm that could be applied to the task of track reconstruction. In particular it neglects the fifth tracking parameter, the impact of multiple scattering, and the advantages that could be obtained by treating the endcap and barrel layers with different parametrisations. This served to simplify the implementation of the KF in the FPGA at the expense of some tracking performance, although it is expected that some extensions could be added without significant redesign of the existing firmware. The first of these is the addition of the transverse impact parameter, d_0 to the fit, which will help identify and improve fitting performance for particles which do not originate from the beam line, such as those from B hadron decay. In this case, the track equation of ϕ would become

$$\phi = \phi_0 - \frac{r}{2R} - \frac{d_0}{r}. \tag{5.18}$$

A fifth parameter would require some additional resources, since the matrix dimensions of $\mathbf{P}_{t|t}$ and $\mathbf{H}_t$ increase with the number of state parameters.

Accounting for the effect of multiple scattering has also been investigated. An additive contribution to the forward prediction of the state covariance matrix, $\mathbf{P}_{t+1|t}$, is required. This is the matrix $\mathbf{Q}_t$, as described in Sect. 5.1.1, and could easily be accommodated within the existing design. Since the magnitude of multiple scattering depends on the amount of material traversed and the particle momentum, some additional computation is required. The resource usage for this is small compared to the other matrix calculations (one additional BRAM per KF worker), and can be executed in parallel. The addition of multiple scattering may improve efficiency by finding additional stubs after scattering, and may improve resolution by de-weighting stubs whose coordinates are more likely to have been influenced by scattering.

The decision to use stub radius as the stepping parameter was motivated primarily by its simplicity to implement in the firmware design. The track equations described in Sect. 5.1.1 naturally suggest using the radius r as the stepping parameter in the KF. This is an appropriate choice for the tracker barrel, where modules are arranged in layers of approximately constant (and precisely known) radius. In the detector endcaps, however, the z coordinate is more precise, therefore stepping in z is a more suitable parametrisation in this region. Since most tracks will pass through modules in the barrel and the endcap, $\hat{x}_{t|t}$ and $\mathbf{P}_{t|t}$ would need to be transformed at the boundary to the new coordinate system (which may increase latency and resource utilisation), and distinct state updater logic would be needed for the update of barrel and endcap states. For a fast and lightweight FPGA implementation of the KF, this would not be desirable, so currently r is used as the stepping parameter throughout, and the uncertainty in r due to the strip length in endcap modules is incorporated into the z uncertainty as $\sigma_z^2 = \sigma_r^2 (\cot \theta)^2$. For tracks with four endcap hits, this degradation in resolution from the $\cot \theta$ factor is between 1.9 and 5.5. An improved precision would improve the ability of the fit to reject bad combinations of stubs in this region. For this reason, the alternative scheme may be used in the final track finder.

It may be desirable to use floating point operations in a future design. As discussed in Sect. 5.1.3, some 35-bit fixed point quantities were used in the design, and this was in part to facilitate a wide range of values. A floating point numerical representation would allow a constant relative precision for these values. MaxJ supports the use of floating point types, including with non-standard bit widths for the exponent and mantissa.

A Boosted Decision Tree (BDT) has been developed in FPGA logic that may be bolted on to the back-end of the KF to select and remove fake tracks that remain following the track fit. A full description of the implementation can be found in [12]. The choice of a gradient BDT with 100 trees, each three nodes deep was motivated by simulation results. The four input parameters are χ^2, $1/p_T$, $\cot \theta$ and the number of tracking layers skipped during fitting. The BDT is trained on a CPU, and the trained parameters are exported to a text file, which can be read into a static FPGA implementation. The BDT can be tuned to increase its fake reduction power at the expense of reduced tracking efficiency caused by the rejection of genuine tracks. It is able to reduce the number of fake tracks in the $t\bar{t}$ with 200 pileup sample by 35, 50, or 70%, with a corresponding loss in efficiency of 0.1, 0.5, or 1.0%. One instance processes one track in 12 clock cycles, and runs at 400 MHz in the Stratix V [16] FPGA. The implementation consists of 9700 LUTs and 9900 FFs per instance.

In addition to these new features, some improvements can be made to the firmware design. A higher clock frequency would decrease the overall fit latency. Some increase beyond the 240 MHz achieved here should be possible, since the architecture of the Kalman filter is such that there are very few fan-outs, which allows connected components to be placed closer in the FPGA, reducing signal path latency. In addition, the design is heavily pipelined which allows for easier clock frequency scaling without breaking the data flow structure.

Although the KF described here is capable of track fitting down to 2 GeV, a discussion of how its performance could be improved at low p_T can be found in Sect. 6.6.

5.4 Seed Filter and Linear Regression Fit

An alternative fitter has been developed that is also able to process track candidates from the HT. This fitter is known as the Seed Filter and Simple Linear Regression (SFLR) [17, 18], and has also been implemented on a Xilinx V7-690. The SFLR first verifies that track candidates from the HT are compatible with a straight line trajectory in the $r - z$ plane. This is the seed filter component. It is able to remove stubs from candidates that prevent the candidate from fulfilling this criteria. The simple linear regression fit then calculates the track parameters of the remaining tracks, with independent straight line fits in both $r - \phi$ and $r - z$.

Just like the KF, the SFLR firmware has been successfully demonstrated in hardware, and the results match the CMSSW emulator well. In comparison with the KF, the SFLR utilises 19% fewer BRAMs, but 29% more LUTs to process the track candidates from one tracker sector (nonant or octant). As tracks are output from the SFLR as they are fitted, and there is no accumulation period necessary, the first tracks are output within 875 ns. The output of the final tracks will be within 1658 ns. These numbers include the latency associated with the duplicate removal algorithm described in Sect. 5.5.1. In terms of performance in simulated $t\bar{t}$ events with 200 pileup, the SFLR outputs 83 (103) tracks in the tilted (flat) geometry scenarios. (To be compared to 73 (79) for the Kalman filter). The final track finding efficiency of the SFLR chain is very similar to the Kalman filter, with 94.3%, however a lower efficiency is observed at $p_T > 20$ GeV in the SFLR case, as it may be less efficient at removing unwanted stubs from track candidates in high p_T jets.

It is unlikely that the SFLR will match the track fitting performance and flexibility of the KF, however it may be a suitable option if latency becomes prioritised in the future. It is yet to be determined which of the two options scales best to higher clock frequencies, and this may become an important factor, as will be discussed in Chap. 7.

5.5 Duplicate Removal

5.5.1 Algorithm

The DR algorithm is the last element in the Track Finding Processor chain. At the input to the DR, over half the track candidates are unwanted duplicate tracks created by the HT. The purpose of the DR is to select and discard these tracks.

The DR algorithm is based on an understanding of how duplicate tracks form within the HT. The chosen granularity at the HT allows stubs to be logged in multiple different cells in the track parameter space. This is illustrated in Fig. 5.6, where in the example shown, four stubs (red lines) associated with a single particle produce three track candidates in the blue and yellow HT cells. The KF rejects incorrect stubs from the tracks and it fits the track parameters with a greater precision, therefore the duplicated tracks now have q/p_T, ϕ_T helix parameters that lie within the same bin as the genuine tracks. In the example in Fig. 5.6, these three tracks contain the same stubs; when they are fitted they will all yield identical fitted track parameters. These fitted parameters should correspond to the blue cell, where the lines intersect. One can therefore compare the original HT cell of each track candidate to the fitted result in $(q/p_T, \phi_T)$ space to determine whether the fitted tracks are likely to be duplicated or not, and discard the duplicates without any need for track to track comparisons.

The DR algorithm can therefore be described as follows: after the track fitter, any track whose fitted parameters do not correspond to the same HT cell as the Hough transform originally found the track in is eliminated. Hence, in the example of Fig. 5.6 the yellow cells will be eliminated and the blue (centre shaded) cell will be kept. The advantage of this algorithm is that it identifies duplicates by looking at individual tracks. As a result, there is no need to compare pairs of tracks in order to find out if they are similar. However, due to resolution effects this simple implementation of

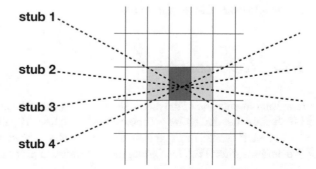

Fig. 5.6 Illustration of the r-ϕ Hough transform showing the formation of duplicates. The dark grey cell represents the desired (genuine) track-candidate, whereas the light grey cells depict duplicate track candidates generated within the HT by the same set of stubs

Fig. 5.7 Demonstration of the duplicate removal algorithm running on a single (highly occupied) HT array in emulation. For an example event, the upper plot shows the occupancy of an HT array after the HT step only. The lower plot shows only the track candidates that remain following the duplicate removal. A single charged particle can generate many adjacent track candidates in the HT array, but these sets of candidates are reduced to just one following duplicate removal

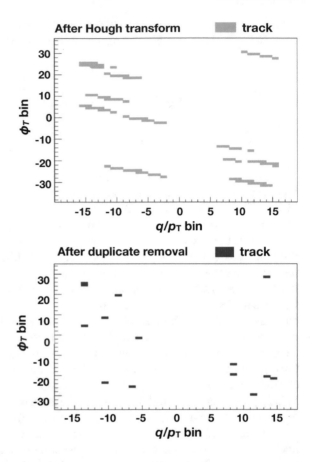

the algorithm loses a few percent tracking efficiency by discarding tracks that are not duplicated. The efficiency can be recovered by performing a second pass through the rejected tracks. During that pass, tracks whose fitted parameters do not correspond to the HT cell of a track that was selected in the first pass are recovered. Figure 5.7 shows the impact of running this duplicate removal algorithm on a highly occupied HT array. This figure shows how lines of track candidate duplicates can be reduced to a single track after duplicate removal.

5.5.2   Implementation

The implementation of the duplicate removal algorithm is shown in Fig. 5.8. The DR block shown in that figure processes the tracks found by the KF in six sub-sectors, so six such DR blocks must be instantiated to process tracks from all 36 sub-sectors in the demonstrator TFP. Designing the DR block to process six sub-sectors instead of

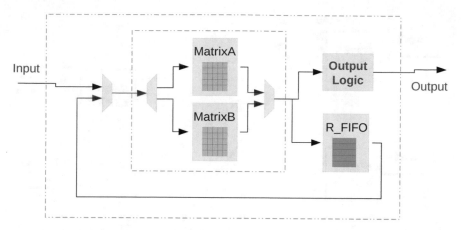

Fig. 5.8 Architecture of the duplicate removal algorithm implementation. A single DR logic block is shown, which processes the KF tracks from six sub-sectors. Therefore, six such blocks are needed to process all 36 sub-sectors in the demonstrator TFP [14]

one minimises the resource usage. As tracks are not ordered or sorted by sub-sector when they are sent from the KF, a router is required to sort each fitted track to the duplicate removal block that handles that sub-sector. The mapping of this router is shown in Fig. 5.9.

Within the DR block, a matrix representing the HT arrays of the six sub-sectors is implemented in a 18 Kb memory, and is addressed using the sub-sector ID and the ID of the $(q/p_T, \phi_T)$ cell that the candidate was found at within the HT array. Any fitted track is flagged as consistent if its fitted helix parameters correspond to the same HT cell as the HT originally found the track in. These tracks are forwarded to the output channel, and the corresponding matrix address is marked. Tracks which are both inconsistent (the helix parameters do not point to the same HT cell that the track candidate was originally found in) and do not correspond to an entry already in the matrix are added to a recovery FIFO (named R_FIFO in Fig. 5.8), which stores these rejected tracks for a final pass.

After all tracks have arrived from the KF, the helix parameters of the inconsistent tracks in the recovery FIFO are checked against the HT cell locations that have been marked in the matrix. If one of the fitted track parameters corresponds to a HT cell not yet marked the track is recovered, forwarded to the output channel, and the corresponding address is marked in the matrix.

A complete reset of the matrix is required before processing tracks from the next event, so two matrices (labelled Matrix A and Matrix B in Fig. 5.8) are instantiated, which take it in turn to process alternate events. There is therefore always one active matrix and one resetting matrix. Along with them, one FIFO is used for each matrix to store the addresses that were marked and hence need to be cleared before the next event. Each matrix and its corresponding clear addressing FIFO together occupy one 36 Kb memory block. The FIFO in which the inconsistent tracks are temporarily

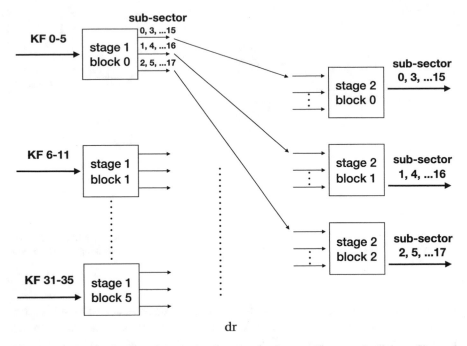

dr

Fig. 5.9 Block-like diagram of the routing firmware implemented between the Kalman filter and the duplicate removal firmware blocks, as implemented in the V7-690 FPGA. The fitted tracks are sorted into three different locations, each one corresponding to six of the eighteen sub-sectors that are processed in a single V7-690

stored uses two 36 Kb block RAMs. Therefore, a total of four 36 Kb block RAMs are used for each DR block, which is designed to process data from six sub-sectors. As shown in Table 5.1, (which lists the resource utilisation for each DR block as it is implemented at 240 MHz within the Xilinx V7-690), this is a very lightweight design when compared to the entirety of the KF. The time to process a track is four clock cycles.

In stand-alone hardware tests, 99.6% of the DR output tracks match exactly the expectations from the software emulator. Around 0.5% efficiency is lost as a result of the DR incorrectly removing genuine track candidates.

The KF also discards tracks that have fitted helix parameters that lie outside the η, ϕ sub-sector that the track was originally found in (during the HT stage). This eliminates a small number of duplicates that exist between the sub-sectors.

References

1. Frühwirth R (1987) Application of Kalman filtering to track and vertex fitting. Nucl Inst Meth A 262:444–450. https://doi.org/10.1016/0168-9002(87)90887-4

2. Kalman RE (1960) A new approach to linear filtering and prediction problems. Trans ASME - J Basic Eng 35–45. https://www.cs.unc.edu/~welch/kalman/media/pdf/Kalman1960.pdf
3. Gaylor DE, Lightsey EG (2003) GPS/INS Kalman filter design for spacecraft operating in the proximity of International Space Station, Aug 2003, AIAA Guidance, Navigation, and Control Conference and Exhibit. https://doi.org/10.2514/6.2003-5445
4. Bar-Shalom Y, Rong LX, Thiagalingam K (2001) Estimation with applications to tracking and navigation. Wiley, New York, pp 308–317. ISBN 978-0-471-41655-5
5. Einicke GA (2006) Optimal and robust noncausal filter formulations. IEEE Trans. Sign Process 54(3):1069–1077. https://doi.org/10.1109/TSP.2005.863042
6. Harvey AC (1994) Applications of the Kalman filter in econometrics. In Bewley T (ed) Advances in econometrics. Cambridge University Press pages, New York, 285f, ISBN 0-521-46726-8
7. Mankel R (2004) Pattern recognition and event reconstruction in particle physics experiments. Rep Prog Phys 67:553. https://doi.org/10.1088/0034-4885/67/4/R03
8. Strandlie A, Wittek W (2006) Derivation of Jacobians for the propagation of covariance matrices of track parameters in homogeneous magnetic fields. Nucl Inst nd Meth A 566(2):687–698. https://doi.org/10.1016/j.nima.2006.07.032
9. Innocente V, Nagy E (1993) Trajectory fit in presence of dense materials. Nucl Inst Meth Res Sect A 324(1–2):297–306
10. Strandlie A, Früwirth (2009) Track and vertex reconstruction: from classical to adaptive methods, Oct 2009. http://www.hephy.at/fileadmin/user_upload/Fachbereiche/ASE/Strandlie.pdf
11. Maxeler Technologies, MaxCompiler white paper, Feb 2011. https://www.maxeler.com/media/documents/MaxelerWhitePaperMaxCompiler.pdf
12. Summers SP (2018) Application of FPGAs to triggering in high energy physics, Mar 2018, Imperial College London Ph.D. Thesis, CERN-THESIS-2018-248
13. Summers SP, Rose A, Sanders P (2016) Using MaxCompiler for high level synthesis of trigger algorithms. JINST 12:C02015. https://doi.org/10.1088/1748-0221/12/02/C02015
14. Aggleton R et al (2017) An FPGA based track finder for the L1 trigger of the CMS experiment at the High Luminosity LHC. JINST 12:P12019. https://doi.org/10.1088/1748-0221/12/12/P12019
15. Xilinx Inc (2017) 7 series FPGAs data sheet: overview, Aug 2017, product specification, DS180 (v2.5)
16. Intel Corporation, Stratix V device overview, Oct 2015, SV51001
17. Cieri D (2018) Development of a L1 Track and Vertex Finder for the Phase II CMS experiment upgrade, Jan 2018, University of Bristol Ph.D. Thesis, CERN-THESIS-2018-045
18. Aggleton R et al (2017) A novel FPGA-based track reconstruction approach for the level-1 trigger of the CMS experiment at CERN, Sep 2017, 27th International Conference on Field Programmable Logic and Applications. https://doi.org/10.23919/FPL.2017.8056825

Chapter 6
Demonstrator Results

6.1 Demonstrator Configuration and Data Format

In order to test the track finding capabilities of the TFP demonstrator and the proposed
final system, simulated physics events with a pileup of up to 200 proton-proton
interactions per bunch crossing were produced with the CMS simulation software,
including modelling of particle interactions with the detector and the generation of
stubs. Samples consisting of top quark pair-production events ($t\bar{t}$) are used primarily
for analysis, as the presence of high energy jets is particularly challenging for the
track finding algorithms.

Software developed to validate the hardware demonstrator is used to inject stubs
from these samples into the demonstrator chain, converting them to a text file before
transmission over IPbus [1]. Tracks reconstructed by the demonstrator using these
stubs are retrieved via IPbus at the end of the chain and are stored for later analysis.
The 48-bit input stub data format is given in Table 6.1. The 128-bit output track data
format is given in Table 6.2.

In parallel with the firmware developments, a software emulation of the TFP
demonstrator has also been developed. The emulator is able to process the same
fixed-point formatted stub data used as input to the demonstrator to produce tracks,
for offline comparison with the hardware output. The emulator uses fixed-point mathematical operations where possible, and attempts to simulate the logic implemented
in the FPGAs in order to model time-dependent effects. However, it is not fully a
clock-cycle accurate emulation so small differences between hardware and emulation are expected. The emulator software can be configured to use full floating-point
precision for comparison. Output tracks produced by the emulator, and those produced by the hardware demonstrator are analysed by a comparison software package,
which checks for consistency on an event-by-event basis. As the TFP firmware is
agnostic to which tracker sector stubs are coming from, it is simple to inject stubs
associated with each sector, one by one, in an iterative manner such that the tracking
performance can be demonstrated for the entire geometrical coverage of the tracker.

© Springer Nature Switzerland AG 2019   89
T. O. James, *A Hardware Track-Trigger for CMS*, Springer Theses,
https://doi.org/10.1007/978-3-030-31934-2_6

Table 6.1 Input stub data format at the start of the demonstrator track finding chain. The resolution of the quantities that are given is not the known uncertainty of the quantities, but rather the maximum theoretical resolution of the fixed-point encoding scheme used. The quantity ϕ_{Oct} denotes the ϕ coordinate of the stub relative to the centre of the ϕ octant. The module type parameter is used to look up the strip pitch, strip length, and sensor separation of the tracker module associated with the stub

Quantity	N. bits	Bitwise resolution
ϕ_{Oct}	15	47.9 μrad
r_T	10	1.56 mm
z	12	1.56 mm
Bend	6	0.25 strips
Module type	3	–
Stub valid	1	–
Reserved	1	–
Total	48	–

Table 6.2 Output data format at the end of the demonstrator track finding chain. The resolution of the parameters given is not the resolution that the parameters are measured to with the demonstrator, but rather the maximum theoretical resolution of the fixed point encoding scheme used; therefore some of the resolution numbers are too precise to be physical. The (excessively) large number of bits used to encode the track parameters was chosen to assist in debugging and validation. Results plots are generated with this truncated to 12 bits. The quantities labelled HT bin correspond to the bin in the HT that the track candidate was originally found in. The quantities labelled helix parameter (param.) bin correspond to the HT bin that the track would be expect to be found in, given its fitted parameters. Number of layers encodes the number of unique tracker layers or endcap disks that contained stubs used to fit the track. The sector ID encodes the η sub-sector that the track was found in

Quantity	# bits	Bitwise resolution
η sub-sector ID	5	1 sector
$1/2R$	18	0.093 km^{-1}
ϕ_0	18	3.00 μrad
$\cot\theta$	18	1.22×10^{-4}
z_0	18	0.491 μm
χ^2	17	7.81×10^{-3}
# of layers	3	–
q/p_T HT bin	5	1 bin
ϕ_T HT bin	6	1 bin
q/p_T helix param. bin	5	1 bin
ϕ_T helix param. bin	6	1 bin
Error and validation	8	–
Reserved	1	–
Total	128	–

As the demonstrator firmware has been developed with the flat barrel geometry (and octant sectors) in mind, all results from the hardware demonstrator are based upon samples generated for this scenario. However, emulation software has been developed that models the changes to the firmware that would be required to perform well in the tilted barrel geometry (and with nonant sectors). Where appropriate, results generated with this software are presented, and should be considered complementary to the hardware (flat barrel) results. As the stub rates in the inner layers are reduced, and the stub generation efficiency is improved, track reconstruction performance with the tilted barrel geometry is generally better than or equal to the flat barrel scenario. For this reason, the satisfactory performance demonstrated in hardware for the flat barrel geometry should be sufficient evidence that achieving the required performance in the proposed tilted barrel, nonant hardware configuration is feasible.

6.2 Tracking Efficiency and Purity

The track reconstruction efficiency is measured relative to all generated charged particles from the primary interaction that produce stubs in at least four layers of the tracker, and are within the defined kinematic acceptance region ($p_T > 3\,\text{GeV}$, $|\eta| < 2.4$, $|z_0| < 30\,\text{cm}$ and $L_{xy} < 1\,\text{cm}$, where L_{xy} is the transverse distance from the beam line to the particle vertex). The (relatively) large value of $L_{xy} < 1\,\text{cm}$, was chosen to capture b-hadron decays, but exclude interactions in the beam pipe. A charged particle is defined to be successfully reconstructed and contributes to the efficiency if the following two conditions are met:

(i) a reconstructed track was fitted from a track candidate (or a sub-set of a candidate) that has stubs produced by the charged particle in a minimum of four tracking layers;

(ii) this reconstructed track was fitted from a track candidate (or a sub-set of a candidate) that consists solely of stubs produced by the charged particle.

Tracks that are not matched with a charged particle are known as misconstructed or fake tracks. If a charged particle matches more than one reconstructed track, subsequent matching tracks are labelled duplicates. After the track fit, all of the tracks satisfying condition (i) must also satisfy condition (ii), as the KF is configured to select only four stubs to fit from each track candidate. Only the first condition is used for the HT tracking efficiency definition, prior to the track fit.

Table 6.3 shows how tracking performance improves through the reconstruction chain for the sample of $t\bar{t}$ events at a pileup of 200. It can be seen that the HT finds tracks with high efficiency, but also reconstructs many fake and duplicate tracks. However, Table 6.3 also shows that the KF eliminates the majority of these fake tracks (but fails to reduce the fraction below 10% in this sample), and the duplicate removal step successfully removes almost all the duplicates.

The mean tracking efficiency (over all applicable $|\eta|$), as measured in hardware is 94.5%, with a mean number of tracks found per event of 80 (20% fakes). These numbers agree to within 99.5% with the results generated by emulation. The agreement,

Table 6.3 Track finding performance on simulated t$\bar{\text{t}}$ events at a pileup of 200, after each stage of the demonstrator chain, in the flat and the tilted geometry. The track finding efficiencies at each stage are listed using the efficiency definitions given in the text, where the HT efficiency requires only condition (i) to be met. Also quoted are the mean numbers of reconstructed tracks per event in the entire tracker, and the subset of these tracks that are unwanted as they are either fake or duplicate tracks

Flat barrel	Efficiency [%]	# of tracks	# of fakes	# of duplicates
HT	97.1	331	139	126
KF	95.1	190	27	103
DR	94.4	79	16	3
Full chain	94.4	79	16	3

Tilted barrel	Efficiency [%]	# of tracks	# of fakes	# of duplicates
HT	97.1	295	104	124
KF	96.3	159	16	84
DR	95.1	73	10	4
Full chain	95.1	73	10	4

when studied as a function of the particle kinematic properties is shown in Fig. 6.1. The tracking efficiency from emulation is almost identical regardless of whether floating-point or fixed-point stub data are used. In the tilted geometry emulation, the efficiency for the same physics sample is 95.1%, with a mean number of tracks of 71 (13% fakes).

As shown in Table 6.4, the efficiency to reconstruct leptons in the t$\bar{\text{t}}$ physics events exceeds 97% for muons over the entire acceptance, but this number is somewhat lower at 81.3% for electrons, as shown in Fig. 6.2. The lower electron efficiency is expected and is mainly due to bremsstrahlung effects, which cause the particle trajectory to deviate from the helix assumed by the tracking algorithms. Some improvement to this lower efficiency should be possible. For example, the KF algorithm can be modified to allow for multiple scattering (see Sect. 5.3). The agreement between hardware and emulation is equally good for leptons.

The purity of track reconstruction in the core of dense jets is slightly degraded in comparison to isolated tracks, due to the increased likelihood of incorrect stubs being included in the track candidate. Figure 6.3 shows that when selecting on charged particles in jets which have total p_T exceeding 100 GeV, there is a small efficiency loss, particularly in the region $|\eta| > 1$. As shown in Fig. 6.3, when adjusting the second matching requirement to allow reconstructed tracks with at most one incorrect stub to contribute to the efficiency this effect is reduced. This indicates that much of the reduction in measured efficiency that is observed in the region $|\eta| > 1$ comes from the reduced track purity in these high-energy jets. Improved rejection of these incorrect stubs by the KF should help improve the overall performance of the track finder.

Fig. 6.1 Track reconstruction efficiency, measured in both hardware and emulation, for tracks originating from the primary interaction in $t\bar{t}$ events with 200 pileup events superimposed, as a function of p_T (upper) and η (lower). Flat barrel geometry

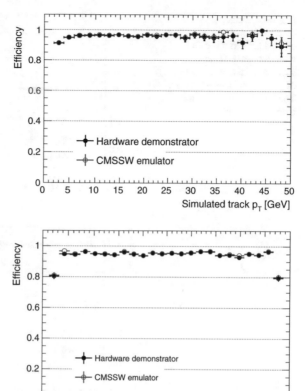

Table 6.4 Performance comparison between flat and tilted barrel tracker geometries using events containing a single muon with fixed $p_T = 10\,\text{GeV}$, at a pileup of 200, reconstructed using the tracking algorithms optimised for a flat barrel geometry. The mean number of tracks found and the tracking efficiency are provided

Single $\mu^{\pm}$ events with 200 pileup	Flat geometry	Tilted geometry
# of tracks after HT	229	161
# of fakes after HT	92	35
# of tracks after full chain	55	48
# of fakes after full chain	9	4
Efficiency after full chain (%)	97.3	97.3

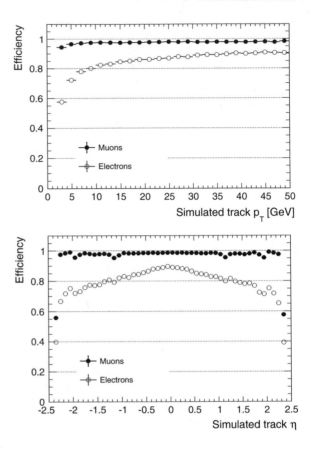

Fig. 6.2 Track reconstruction efficiency, for $e^{\pm}$ and $\mu^{\pm}$ from $t\bar{t}$ events with 200 pileup events superimposed, as a function of p_T (upper) and η (lower). These results are obtained from emulation

The misconstructed track (fake) rates in $t\bar{t}$ and 200 pileup events, as a function of reconstructed η and q/p_T are shown in Fig. 6.4. The high fake rate around $|\eta| = 1.5$ is a result of a higher stub-on-track multiplicity in this region. The ratio of the number of fake tracks over the number of genuine tracks is given as a function of $1/p_T$. The large fraction of fake tracks at very high p_T is a result of very few genuine particles having such high p_T, whereas the distribution of combinatorial fakes is flatter by comparison.

6.3   Track Parameter Resolution

Figure 6.5 shows the resolution of the four track parameters (p_T, ϕ_0, $\cot\theta$, z_0) for reconstructed primary tracks from both hardware and emulation in $t\bar{t}$ events with a pileup of 200. A fifth track parameter, d_0 is not reconstructed by the KF, but there is potential to add this functionality (see Sect. 5.3 for further discussion). The resolution is defined as the rms of the residuals between the measured track parameter and

Fig. 6.3 Track
reconstruction efficiency as a
function of η, for all tracks
originating from the primary
interaction (black dots), or
for only the tracks contained
within a primary jet that has
a total p_T exceeding
100 GeV (red open circles),
for $t\bar{t}$ events with 200 pileup
events superimposed. Either
no incorrect stubs are
allowed on the track (upper),
or at most one incorrect stub
is allowed on the track
(lower). These results are
obtained from emulation of
the flat barrel geometry

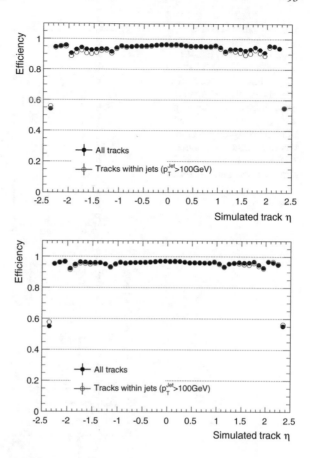

the simulated truth track parameter. The level of agreement between hardware and
emulation is as expected, with remaining differences due to the use of floating-point
arithmetic in parts of the emulator code. A degradation in resolution with increasing
η is observed, due to a combination of the shorter effective lever arm available,
the reduced effective precision for hits in the endcap and the impact of increasing
material traversed by particles (and therefore increased scattering).

It is also instructive to compare the resolution of parameters for particles with
different p_T ranges. The resolution of the track parameters for single isolated muons
is provided in Fig. 6.6. Multiple scattering effects dominate at low p_T, and this is
particularly evident in the ϕ_0 resolution (Fig. 6.6), where the resolution is better than
0.4 mrad for muons with $15 < p_T < 100$ GeV, and between 0.7 and 1.5 mrad for
muons with $3 < p_T < 5$ GeV. Similar effects are observed in the $\cot\theta$ resolution,
where a resolution of around 0.0025 for $p_T < 5$ GeV tracks that only pass through
the barrel is observed, but this degrades by approximately a factor of ten for tracks
at the highest η ranges. The relative precision in p_T for muons with $p_T < 15$ GeV
is limited by multiple scattering, and is approximately 1% in the barrel. This metric

Fig. 6.4 Misconstructed track (fake) rate as a function of reconstructed η (upper) and q/p_T (middle). The ratio of the number of fake tracks over the number of genuine tracks is given as a function of $1/p_T$ (lower). The sample of $t\bar{t}$ and 200 pileup is used. On average there are about 10 fakes per event, out of about 70 reconstructed tracks

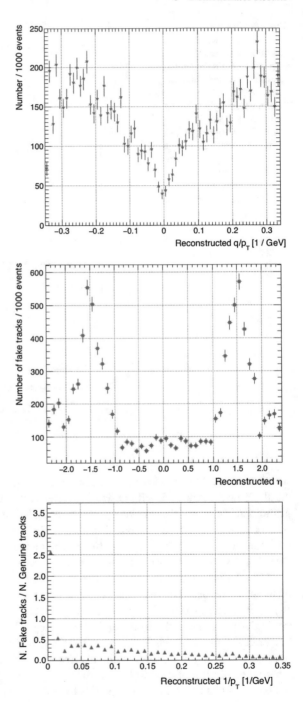

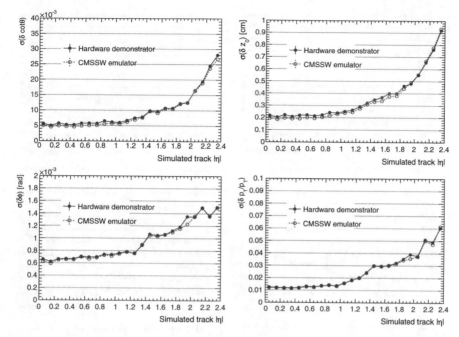

Fig. 6.5 $\cot\theta$ resolution (upper left), z_0 resolution (upper right), ϕ_0 resolution (lower left), and relative p_T resolution (lower right), measured for tracks originating from the primary interaction in $t\bar{t}$ events with a pileup of 200 as determined from both hardware and emulation. Flat barrel geometry

actually degrades with increasing p_T due to decreasing radius of curvature, and is about 2% for barrel only tracks at high p_T.

The resolutions compare reasonably well to those obtained with offline track reconstruction [2], and are good enough to ensure that the tracks will be useful to the L1 trigger [3]. It should be noted that the offline reconstruction is able to utilise all available information from the tracking system, and more sophisticated reconstruction algorithms. In offline simulations, for 10 GeV muons passing through the centre of the tracker barrel, the ϕ_0 resolution is approximately 0.2 mrad while the p_T resolution is approximately 0.5%. Full offline reconstruction outperforms the demonstrator by about an order of magnitude in the z_0 and $\cot\theta$ fit, primarily due to the inclusion of hits from the inner pixel detector.

By comparing the z_0 and $\cot\theta$ resolution of single isolated muons with fully floating point simulations, as shown in Fig. 6.6, the demonstrator only achieves approximately half of the floating point precision in the barrel. This degradation of resolution in the demonstrator system comes as a result of choosing to encode the r and z stub coordinates too coarsely (using 10 and 12 bits, respectively) and therefore reducing the ultimate precision of the track parameters. Figure 6.6 shows that improving the encoding of the stub coordinates by assigning an additional two extra bits to both r and z recovers the lost precision, meaning that the z_0 resolution reaches approximately 1 mm for muons with $5 < p_T < 15$ GeV and $|\eta| < 2$. As there is one reserved

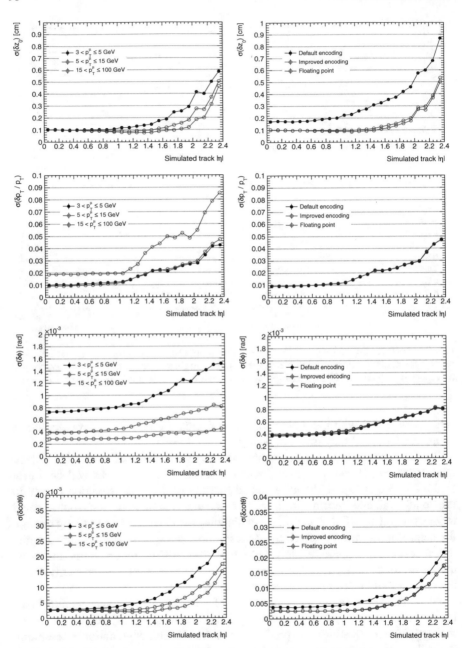

Fig. 6.6 z_0 (top row), relative p_T (second from top row), $\cot\theta$ (second from bottom row), and ϕ_0 (bottom row) resolution, measured for single isolated muons (left column) with $3 < p_T < 5$ GeV, $5 < p_T < 15$ GeV, and $15 < p_T < 100$ GeV. Resolution for muons over the full $3 - 100$ GeV range are also shown (right column), with default (10-bit r, 12-bit z), enhanced (12-bit r, 14-bit z), and full floating point input stub precision. These results are obtained from emulation. In the figure, p_T^μ simply means the p_T of the generated muon

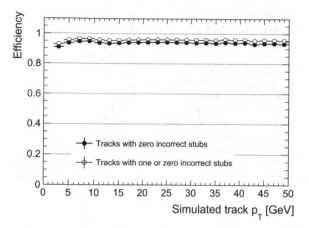

Fig. 6.7 Efficiency of track finding in jets as a function of p_T, when one incorrect stub is allowed, and when none are allowed. The tracks that are used were selected to originate from the primary vertex of a simulated top quark decay. These results were obtained from emulation with flat barrel geometry

bit in the input stub data format, and the three bits used to code the module type are not required and can be replaced with a look-up-table in the HTP, the total number of bits used to encode the stub can remain at 48. This change can be implemented with negligible impact on resource usage, and no impact on latency. Figure 6.6 also shows that with this improved encoding scheme, the precision of all four parameters is close to that obtained by the floating-point simulation of the demonstrator. This effect is not observed in the p_T and ϕ_0 resolutions, as the z-coordinate is not used to fit these parameters.

An incorrect stub on a track candidate can be defined as one that was not created by the same particle as the majority of the stubs that contributed to the track candidate. Stubs that were generated by multiple particles (e.g different particles generated the cluster in the upper and lower sensor) are counted as correct if any one of those particles is the particle of interest. If tracks found by the KF with one incorrect stub were included in the efficiency numerator, the efficiency for finding tracks in jets would be slightly higher (particularly for $|\eta| > 1.1$, and at high p_T). This is shown in Fig. 6.7. However, in general these additional tracks have much poorer track parameter resolution, particularly in the z_0 parameter. A single incorrect stub can reduce the precision of the z_0 fit from $\sim$2 mm to $\sim$2 cm. As such tracks would be of much lower value to the L1 trigger [3, 4], it is therefore required that no incorrect stubs may be used to fit a track candidate if it is to be matched to a particle for the purposes of efficiency measurements.

6.4 Data Rates and Limitations

As shown in Fig. 6.8, the number of tracks reconstructed per event increases with increasing pile-up. On average, 73 (79) tracks per event are reconstructed in $t\bar{t}$ events with a pileup of 200 in the tilted (flat) geometry.

Fig. 6.8 Total number of reconstructed tracks per event when processing $t\bar{t}$ events superimposed with 0, 140, and 200 pileup events. These results are obtained from emulation, and are shown when effects of truncation, caused by excess data flow through the system, are both included and excluded. Flat barrel geometry

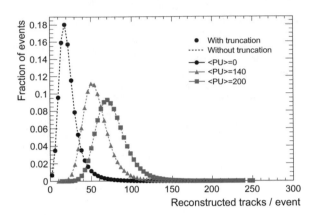

The demonstrator system has been designed to handle the high data rates present in these events with minimal data loss or truncation that results when the firmware does not have enough time to complete the processing of one event prior to data from the next event becoming available. Figures 6.9 and 6.10 show the distribution of the number of stubs per event transmitted from the HTP to the HT in each sub-sector. Truncation of data would occur if these stubs could not all be sent within the 900 ns period before the next event arrives (as is specified by the 36 bx demonstrator time multiplexing period). As one stub from each sub-sector is transmitted at 240 MHz, this corresponds to a theoretical limit of 216 stubs per sub-sector, although data truncation occurs from around 175 stubs due to gaps in the data sequence. This limit exceeds the average data rate by a factor of 1.94 (2.92) in the flat (tilted) barrel geometry. Truncation effects in this part of the system are small, but measurable: in the flat barrel scenario 0.3% of stubs are lost in $t\bar{t}$ events at pileup of 200, which in turn leads to a 0.5% loss of tracking efficiency in the flat geometry.

Figures 6.9 and 6.10 also show the distribution in the number of reconstructed tracks in each sub-sector that are output by the HT per event, for the same sample. It should be noted that 70% of sub-sectors contain no reconstructed tracks, while 97.5% of sub-sectors contain fewer than ten reconstructed tracks. As the mean number of stubs associated with each HT candidate is 6.8, there is usually no difficulty in outputting tracks from the HT within the latency limitation. The effects of this limitation are most evident when collimated, high-energy jets from the $t\bar{t}$ system produce many particles and stubs in a narrow angular region (e.g contained within one or two ϕ, η sub-sectors), accounting for the tails seen in Figs. 6.9 and 6.10. The array division and subsequent multiplexing and load balancing introduced at the back-end of the HT, as discussed in Sect. 4.2.3, address this challenge. With these modifications to the HT implementation, the loss in tracking efficiency due to truncation at the output of the HT is below 0.1% when processing $t\bar{t}$ events with a pileup of 200 and the flat geometry.

As described in Sect. 5.1.4, the accumulation latency of the KF track fitter is configured such that it has time to assign stubs from four tracker layers to almost all

Fig. 6.9 Data rates out of the HTP and into the HT, when processing $t\bar{t}$ events with 200 pileup in the flat barrel geometry. The upper plot show the number of stubs transmitted from the HTP to the HT per sub-sector per event. Truncation effects occur when this number exceeds about 175. The lower plot show the number of reconstructed tracks from the HT per sub-sector per event. In the TFP demonstrator design, truncation occurs when the number of stubs on track candidates exceeds 202 (which corresponds to around 30 tracks on average)

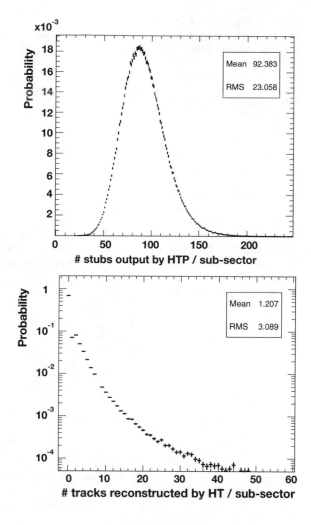

tracks. When processing $t\bar{t}$ events at pileup of 200 less than 0.1% of efficiency is lost in the KF when selecting on these four stub tracks. Again, this loss occurs mainly within particularly high-energy jets, and can be partly mitigated by prioritising the reconstruction of high p_T track candidates.

The total loss in tracking efficiency from truncation effects of the full tracking chain is determined to be less than 0.6% when processing $t\bar{t}$ events at a pileup of 200 in the flat barrel geometry. This can be seen in Fig. 6.11.

Fig. 6.10 Data rates out of the HTP and into the HT, when processing tt̄ events with 200 pileup in the tilted barrel geometry. The upper plot show the number of stubs transmitted from the HTP to the HT per sub-sector per event. Truncation effects occur when this number exceeds about 175. The lower plot show the number of reconstructed tracks from the HT per sub-sector per event. In the TFP demonstrator design, truncation occurs when the number of stubs on track candidates exceeds 202 (which corresponds to around 30 tracks on average)

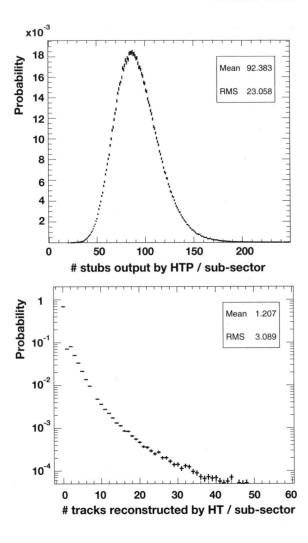

6.5 Tracking Robustness

The demonstrator was also tested on samples that emulated a situation in which a fraction of modules in the tracker do not produce stubs (in the flat barrel geometry). Figure 6.12 illustrates the localised loss in efficiency expected when all modules on barrel layer four, in the region $-1 < \eta < 0$ and $0 < \phi < \pi$, are prevented from generating stubs in simulation. As shown in Fig. 6.12, this efficiency loss can be recovered by relaxing the threshold criterion on the number of hit layers in the HT from five to four in the affected (η, φ) sub-sectors. This change in the threshold leads to a small increase in data rate out of the HT, to 347 track candidates per event (304

Fig. 6.11 Track
reconstruction efficiency for
tracks originating from the
primary interaction in $t\bar{t}$
events with 200 pileup, as a
function of p_T (upper) and η
(lower). It can be seen that
the truncation effects are
negligible. These results are
obtained from emulation,
both including and excluding
truncation effects, in the flat
barrel geometry

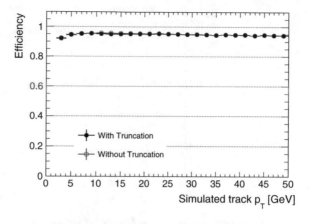

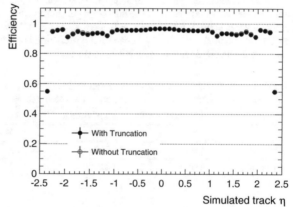

without the recovery, but with the dead modules). The small increase is caused by
additional fake tracks, which occur as a result of the looser threshold.

6.6 Track Finding Down to 2 GeV

Tracking information down to 2 GeV may be of use to the L1 trigger, and the impact
of this requirement on the proposed track-finder system has been studied. Lowering
the minimum p_T threshold from 3 to 2 GeV requires modifying the HTP parameters
to ensure adequate duplication in ϕ, and modifying the HT configuration by increasing the number of columns along the q/p_T axis by 50% to take into account the
increased p_T range, while preserving the precision of the estimate. This increase in
the q/p_T range consequently increases the required FPGA resources by around 50%,
as the resource usage is approximately proportional to the number of columns in the
HT array. When this naive scaling is applied, a reduction of tracking efficiency is ob-

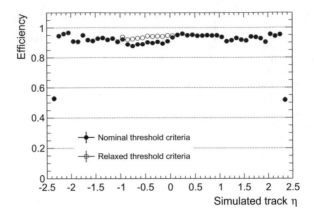

Fig. 6.12 Track reconstruction efficiency as a function of η, measured in emulation, when processing $t\bar{t}$ events with 200 pileup, in the flat barrel geometry, where the tracker is affected by a failure of all modules in the region $-1 < \eta < 0$ and $0 < \phi < \pi$ of barrel layer four. Results are compared before (black dots) and after (red open circles) relaxing the threshold criterion on number of hit layers in the affected region, as described in the text

Table 6.5 Track finding performance down to 2 GeV, for simulated $t\bar{t}$ events at a pileup of 200, after each stage of the demonstrator chain, in the tilted geometry

Tilted barrel	Efficiency [%]	# of tracks	# of fakes	# of duplicates
HT	94.8	771	201	381
KF & DR	91.6	210	29	3
Full chain	91.6	210	29	3

served in the range $2 < |p_T| < 2.7\,\text{GeV}$ when compared to the $|p_T| > 3\,\text{GeV}$ range, primarily due to multiple scattering (which can be seen in simulation) where stubs do not always intersect within a single HT cell, and therefore fail to exceed the threshold criteria and generate track candidates. To mitigate this effect, it is possible to reduce the precision of the HT along q/p_T and ϕ_T, for the range $2 < |p_T| < 2.7\,\text{GeV}$ only, by a factor of two. This variable precision HT has been implemented in firmware, and has been successfully demonstrated in hardware. Overall tracking efficiency, rate, and purity for this configuration is shown in Table 6.5.

This implementation results in a track-finding efficiency post KF (HT) of around 87% (92%) in the region $2.15 < |p_T| < 3\,\text{GeV}$, and around 79% (85%) in the region $2 < |p_T| < 2.15\,\text{GeV}$ for $t\bar{t}$ with 200 pileup samples in the tilted geometry. As the KF has not been re-optimised for the new minimum p_T threshold, around 5% efficiency is lost for $p_T < 3\,\text{GeV}$ tracks. This is likely to be improved if the χ^2 cuts are loosened for low p_T tracks as the mean track χ^2 is almost four times larger for 2 GeV tracks in comparison to 8 GeV and above. This is a result of a combination of increased multiple scattering at low p_T, and missing error terms for the inner barrel modules that are tilted in the $r - z$ plane by an angle, γ, not yet included in the fit. The

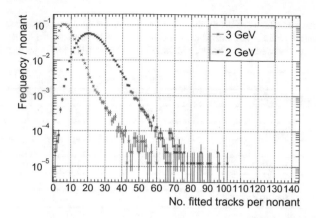

Fig. 6.13 Track rates per nonant out of the TFP, with the tilted geometry, t̄t and 200 pileup. Results are shown for tracking down to 3 GeV, (red) and down to 2 GeV (blue). The vertical axis represents frequency/nonant/event, and is logarithmic. This metric is important for the L1 trigger which must be able to accommodate the tails in the distribution. It is likely that two 16 or 25 Gb/s links per track finder will send data to the L1 correlator

additional errors in ϕ are proportional to $0.0015 \sin \gamma / p_T$, where 0.0015 is the width of a PS module macro-pixel in metres.

Figure 6.13 shows the distribution of track rates per nonant when tracking down to either 2 or 3 GeV. The mean rates per nonant are 23 (2 GeV) and 8 (3 GeV) respectively.

6.7   Latency Measurements

Latency measurements were made for each step independently, and also for the full demonstrator chain. These are shown in Table 6.6, and diagrammatically in Fig. 6.14. Measurements include optical transmission delays and SERDES latency of the links. The full demonstrator chain from source to sink, and the sum of each step measured individually give identical results. By design, the processing latency of the complete system is fixed, regardless of pileup or occupancy. This is done to guarantee delivery of the first and last tracks from an event to the downstream processing in a predictable way. In addition to the time difference between the first stub entering the system and the first track leaving it (FIFO), the table also shows the time difference between the first stub entering and the last track leaving (FILO). Explanation of the latency for each processing step is given in Chaps. 4 and 5. As a result of the time-efficient data delivery scheme to the TFP that is employed, it can be seen that the majority of the latency budget is used in executing the track-finding and track-fitting algorithms. A discussion of how these latency numbers could be optimised in the future, and how they scale to a proposed final system is given in Sect. 7.1.

Table 6.6 Measured latency of the hardware demonstrator, with demonstrated components of the track reconstruction chain listed individually and accumulated, including the SERDES and optical transmission delays between each board. The labelling of the SERDES stages matches that given in Fig. 6.14

Step	Latency [ns]
SERDES + optical length 1	143
HTP	251
SERDES + optical length 2	144
HT	1025
SERDES + optical length 3	129
KF + DR	1658
SERDES + optical length 4	129
Last out − First out	225
Total: First out − First in	3479
Total: Last out − First in	3704

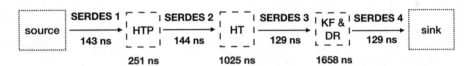

Fig. 6.14 Illustration of the demonstrator latency. The processing latency for each algorithm block is shown, in addition to the latency measured for each stage of SERDES and optical lengths. The labelling of the SERDES stages matches that given in Table 6.6

6.8 The Evolution of the Track Finder

So far the majority of results that have been presented show the performance of the demonstrator and proposed TFP, as it currently stands. It is important to note that it was a 3 − 4 year long process of research, development, optimisation, and trial and error before a system compatible with the L1 requirements was achieved. Table 6.7 shows the state of the proposed track fitter at different points in time during the PhD. May and Dec. 2016 correspond to the Stage 1 and Stage 2 demonstrators respectively, as described in Sect. 3.3.1.

It can be seen in Table 6.7 that one of the most significant improvements over time was the number of track candidates (of which most were not genuine) produced in the HT. An early design (Dec. 2014) generated 3600 candidates from a sample of t$\bar{\text{t}}$ and 140 pileup. For comparison by May 2016 increased η segmentation, the addition of a bend filter, and an optimised offset in the HT array resulted in the same sample (t$\bar{\text{t}}$ and 140 pileup) producing only 275 track candidates on average.

Until Summer 2016, there was no working track fitting or duplicate removal firmware in the demonstrator chain. It was clear that this would be required, but the track fitter must be able to remove stubs from tracks and cope with multiple

Table 6.7 The evolution of track finding components and their configuration and performance, Dec. 2014 to Dec. 2017. The efficiencies given are the HT tracking efficiency up until May. 2016, after which the full definition is used. The physics samples used for performance comparisons are $t\bar{t}$ with either 140 (upper table) or 200 (lower table) pileup. Large samples of 200 pileup events were not available until May 2016. The subscript in the segmentation rows refers to the values of T for ϕ, and S for η respectively

140 pileup	Dec. 2014	May 2015	Dec. 2015
HT firmware	systolic	systolic	systolic/pipelined
Track fit firmware	–	–	–
Geometry	Flat	Flat	Flat
ϕ segmentation	64 in ϕ_0	64 in ϕ_0	32 in ϕ_{45}
η segmentation	5 in η_0	5 in η_0	5 in η_{65}
HT size (ϕ, q/p_T)	32 × 32	32 × 32	32 × 32
HT offset, T	0	65	45
Bend filter	✗	✗	✓
Efficiency [%]	97	96	97
# tracks	3600	2400	1050
200 pileup	May 2016	Dec. 2016	Dec. 2017
HT firmware	HTP & daisy chain	HTP & daisy chain	HTP & daisy chain
Track fit firmware	–	KF & DR	KF/SFLR & DR
Geometry	Flat	Flat	Tilted
ϕ segmentation	32 in ϕ_{58}	16 in ϕ_{58}	18 in ϕ_{61}
η segmentation	9 in η_{65}	18 in η_{50}	18 in η_{50}
HT size (ϕ, q/p_T)	32 × 32	64 × 32	64 × 32
HT offset, T	58	58	61.2
Bend filter	✓	✓	✓
Efficiency [%]	97	94	95
# tracks	800	79	73

stubs on a single tracking layer, which are not features of many standard track fitting algorithms. A linear χ^2 fit, solved by Newtonian iteration; and the Kalman filter; were both studied in software to compare performance. Implementing the KF in firmware was challenging. Bugs in the complex state updater, relating to the number of bits used for intermediate calculations, and overflowing outputs were difficult to diagnose.

A significant amount of time was required to debug the firmware and emulation software to a state where the results were in agreement (to the 0.1% level). While development of the firmware in Xilinx software often proceeded without hitches, the firmware did not always behave as expected when run in the MP7. Differences between firmware simulation in the Xilinx software, and the output of the MP7 hardware were particularly difficult to debug, and involved inserting an ila core into the firmware, which allows the study of internal digital signals in real time with software known as ChipScope Pro [5]. These differences occur because it is not possible to simulate all aspects of the design simultaneously (in particular the infrastructure), and the simulation is very slow, meaning that running a large number of samples through the hardware is often the quickest and easiest way to test for and diagnose bugs. A significant amount of work was also required to integrate all the firmware blocks together to make the demonstrator chain. Subtle differences in timing, data formatting and packaging, or other small bugs often prevented firmware blocks from correctly communicating with each other.

Since Dec. 2016 work has focussed on resolving remaining firmware bugs, optimising and adjusting the algorithms and firmware for the tilted barrel geometry, running the firmware implementations at >240 MHz clock speeds, and developing firmware and hardware to utilise the latest generation of FPGAs and optical links (see Chap. 7).

6.8.1   Rejected Ideas

The concept of an η filter within the Hough transform was investigated. A simple η filter was able to reduce the number of candidates to 580 in the May 2015 configuration, with a corresponding 2% efficiency loss. The concept was to bin stubs from each selected HT cell into 64 (overlapping) bins in η. The mean η of the stubs within the most populated bin would be calculated, and stubs in the candidate more than the calculated resolution away from the mean value would be rejected. Either this filter would take place before the number of layers criterion was applied, or it would need to be applied a second time.

There were plans to run a second HT in the $r - z$ plane. This HT would reduce the number of candidates in the May 2015 configuration to about 400, without any efficiency loss. Although the $r - z$ HT was able to reduce the fraction of fake tracks from 77 to 21%, it produced new duplicates, increasing the genuine duplicate rate from 12 to 69%. The FPGA resources required to run a second HT were ultimately deemed to be too great.

Both the $r - z$ and the η filter were ultimately made unnecessary, as the increase in the number of η sub-sectors from 5 to 9 led to a reduction in fake track candidates by a factor of three. However, these ideas eventually led to the $r - z$ seed filter, as described in Sect. 5.5.1.

An alternative duplicate removal algorithm was studied in software, and implemented in a V7-690. This algorithm compared nearby track candidates in the HT array, and discarded tracks that shared a high fraction of stubs with another candidate. The latency to complete this procedure was in excess of 900 ns, primarily because the DR must wait for all the HT tracks to be read out. As the simple duplicate removal algorithm discussed in Sect. 5.5.1 takes around 38 ns, and performs similarly, it was selected instead.

6.9 FPGA Resource Usage

Constructing the hardware demonstrator out of many MP7 boards avoids the logic constraints of a single, currently available, FPGA processing board. On the other hand, it is important to keep the total resource usage realistic such that a final system could be built at a reasonable cost, using FPGAs that are expected to be available on the time-scale for production. Table 6.8 shows the total FPGA resource usage for each demonstrator step (where the numbers given for the HT and KF implementations are summed across the two boards used for each component). The combined total for all the steps gives the resources used to demonstrate the functionality of one entire TFP with a time-multiplexed factor of 36. Each FPGA in the demonstrator also runs the MP7 core infrastructure firmware (Sect. 3.3.2.1), which is required for board configuration, link buffering and error checking. This firmware was developed for the CMS calorimeter trigger, and while it does not constitute the majority of logic in the TFP (as shown in Table 6.8), it is expected that with some optimisation it could be reduced in size and still deliver the functionality needed for the track finder. Resources required to buffer and transmit track finder data to the DAQ is not included.

In order to meet timing and routing constraints in the V7-690, the designs often prioritised the use of block RAM over LUT-based distributed memory. This balance could be readdressed in the future, as the design is adapted to newer FPGAs, with different types of embedded memory, such as the newly developed UltraRAM [7], a dual-port 288 Kb (larger than the 36 Kb BRAM currently utilised) on board memory block, that can be chained together to create large memory arrays up to 500 Mb in size.

Table 6.8 Total resource usage for the demonstrator TFP (with time-multiplexed factor of 36), as implemented in the V7-690 FPGA [6]. The resources needed to construct a complete TFP correspond to the sum of the numbers from the three rows labelled HTP, HT, and KF and DR. Resources required to buffer and transmit track finder data to the DAQ is not included. Other FPGAs are shown for comparison

Firmware	LUT$[10^3]$	DSP	FF$[10^3]$	BRAM
HTP	121	1056	205	222
HT	244	2304	299	1188
KF and DR	432	5112	366	2008
Infra. per MP7	90	0	91	291
TFP Total (excl. infra.)	795	8472	870	3392
TFP Total (inc. infra)	1245	8472	1325	4857
Available in V7-690	433	3600	866	1470
Available in KU-115	633	5520	1266	2160
Available in VU-9P	1182	6840	2364	2160
Available in VU-11P	1296	9216	2592	2016

6.10   The Associative Memory Track Finder

In contrast with the all-FPGA track candidate finder presented in this thesis, a common alternative approach to tracking within a sub-millisecond latency is to use a custom pattern recognition ASIC, known as an Associative Memory (AM) chip.

A current example where an AM chip is to be used is in the ATLAS FTK system [8, 9], which has been designed for online track finding at up to 100 kHz in real time. The 65 nm CMOS AM06 chip used in this system contains memory banks that store data organised in 18-bit words, where a collection of eight such words is known as a pattern. This chip is capable of storing about 130,000 patterns (spread across 64 blocks with 2000 patterns each), and has an input data bandwidth of about 2 Gb/s, and an output bandwidth of about 0.2 Gb/s. The full FTK system will be composed of 7500 AM chips.

While suitable for the FTK application, the AM06 chips would not meet the requirements for use at CMS Phase II, where we require both operation at an event rate of 40 MHz, pattern recognition on the time-scale of 1 μs, and with a pileup of 140 or 200. The AM06 chips are designed for higher latency applications, and the pattern density and output bandwidth are not sufficient. To this end, development is ongoing for a 28 nm AM chip [9], which would have a pattern density four times greater than the AM06. This chip is designed for a 200 MHz clock frequency, and is planned to consist of four cores with separate outputs, for increased output bandwidth. Further iterations of this chip are planned, with a potential candidate for the CMS track finder expected around 2020.

Similarly to the TFP demonstrator described in this thesis, an AM-based track finding demonstrator has also been developed, although in a less complete form [2, 10]. This proposal makes use of a data delivery layer, an AM pattern recognition layer, and lastly an FPGA-based track fitting step. An FPGA is also needed to distribute the data to up to 16 AMs per prm board. This demonstrator assumes AM ASICs will achieve the specifications of the current R&D design goals, corresponding to about 150,000 patterns per chip. In order to demonstrate the low latency functionality of the future chip, an FPGA emulation of the proposed AM was used, with a reduced pattern capacity (1024 patterns). The final chip will be required to store a few hundred thousand patterns per chip, for a total of around one million patterns for each trigger tower (of which there are 48).

The proposed data processing sequence is heavily pipelined, assumes a 20 bunch crossing time multiplexed period (corresponding to an event rate of 2 MHz), and proceeds as follows:

- Stub data enters the prm. The full resolution stub is stored in a buffer, while a coarse resolution version of the stub (called a *super-strip*) is generated and transmitted to the AM chips;
- The AMs identify sets of super-strips that match patterns called *roads*. The roads are transmitted back to the FPGA;

- Super-strips associated with the matched roads are retrieved from a pattern memory, and the full stub data associated with those super-strips are retrieved from the buffer;
- The super-strips selected are filtered so that each track candidate contains no more than one stub for each tracking layer;
- The track candidate is sent to the track fitter which calculates the track helix parameters. The linearised track fit is implemented in FPGA firmware and uses a simplified Principal Component Analysis (PCA) procedure [10].

In this system, the latency for the first track candidate to be ready for downstream filtering is 1850 ns (with the final candidate appearing at 2300 ns). The majority of this time, however, is taken up in the data delivery, as the time from the last stub in to the first stub out is only six clock cycles. The total latency is very similar to the 1813 ns that is observed for the first track candidate to emerge from the Hough transform (240 MHz version) in the TFP demonstrator. Here an estimate of 250 ns is used for the latency of the DTC stub processing and time multiplexing, and about 150 ns are required for SERDES and optical transmission time between the DTC and HTP, and between the HTP and the HT respectively.

The AM performance has been verified in a flat barrel simulation for scenarios up to a pileup of 250 events. The track reconstruction efficiency of muons above 3 GeV in $t\bar{t}$ events is above 95%, with a sharp turn on curve at 3 GeV. For jet p_T greater than 100 GeV, a reduction of tracking efficiency by up to 35% is observed as a result of high occupancy [2]. This loss primarily affects the lower p_T tracks as the readout of high p_T roads is prioritised. In this simulation, truncation occurs when the number of AM roads exceeds 250, and the number of track stub combinations exceeds 800 per tower. In $t\bar{t}$ events at 200 pileup (flat barrel geometry), truncation due to these limitations is observed to a greater degree than what is observed for the daisy chain HT implementation with the multiplexer.

Once again, the ordering of tracks by p_T can go some way to mitigate the effect that this will have on trigger performance, in addition to future developments in cleaning and fitting techniques, and the change to the lower occupancy tilted barrel geometry. Following the track fitting step, which has an estimated latency of 0.5–1.5 μs, depending on requirements, and the duplicate removal (which is estimated to take an additional 100 ns), a mean track rate for this sample is observed to be $63 - 77$ tracks per event, depending on the choice of combination builder (discussed in [2]). This compares very well to the results from the full TFP demonstrator chain presented in this paper, in which a mean of 79 tracks per event are observed. The additional tracks in the TFP demonstrator are misconstructed (fake) tracks, which are present as a result of looser cuts and less data truncation, designed to maximise efficiency in this approach.

While the all-FPGA demonstrator has been shown to operate using presently existing hardware, an AM chip that can fulfil the requirements of CMS does not currently exist. The cost and time required to develop a chip that has both the latency and memory requirements would likely be considerably more than designing data processing boards with the latest commercially available FPGAs. The all-FPGA demonstrator presented in this thesis contains the full functionality of the final system, and a track finder could be assembled with 5 (number of MP7s used to demonstrate one TFP)$\times$9 (number of ϕ-sectors) $\times$ 36 (demonstrated time multiplexing period)$= 1620$ MP7 boards. Although this would be entirely unnecessary, and would result in an more expensive and larger system than ideal, it does show that an all-FPGA solution is overall a very low risk option. On the other hand, the AM demonstrator relies heavily on the scaling up of the FPGA emulations, which themselves are not fully sufficient to demonstrate the required chip functionality. Considering this, and the cost and risks associated with developing an unproven custom ASIC, in an unfamiliar (for HEP) 28 nm technology, it is expected that an all-FPGA track-finding approach will be used by CMS.

6.11   The Tracklet Track Finder

In addition to the AM and the Hough-transform methods discussed in this thesis, a third track finding system has been proposed and studied. This is an all-FPGA solution known as the *tracklet* approach [11, 12].

The tracklet algorithm is as follows:

- Seeds known as tracklets are formed from pairs of stubs in adjacent tracking layers, and consist of an estimate of the potential track parameters using the interaction point as an additional constraint. A finite number of seeding layer combinations are implemented. The five combinations are barrel layers (1, 2), (3, 4), (5, 6) and endcap disks (1, 2) and (3, 4) either side;
- The tracklets are projected to the other layers (both towards and away from the interaction point), and consistent stubs are included in the track candidate;
- A linearised χ^2 fit is applied to all the stubs in the candidate;
- Duplicates are removed by comparing pairs of track candidates for common stubs.

This approach uses 28 ϕ-sectors, and a time-multiplexed factor of 6. These values are subject to change if tracklet seeding is selected for the final system. Each ϕ sector is further sub-divided into virtual modules whereby only a small fraction of virtual module pairs are consistent with the tracks of interest. At 140 pileup, around 20 tracklets are produced per ϕ sector, per seeding layer combination.

When truncation and digitisation effects (firmware limitations) are not included, the track finding efficiency, rates, and track parameter resolutions of this approach are very similar to that described for the demonstrator chain in this thesis. However, due to combinatorial-heavy stages of the algorithm, the firmware does not have the time to process all the stubs if a virtual module contains a dense jet, as can be found in t$\bar{\text{t}}$ samples. This effect causes a reduction in overall tracking efficiency by about 10% at $|\eta| < 1$, and up to 20% at $|\eta| > 1$, for t$\bar{\text{t}}$ samples at 200 pileup. Without pileup (t$\bar{\text{t}}$ events), a similar efficiency loss is observed at $|\eta| > 1$, but only a few percent is lost at $|\eta| < 1$. Work is in progress to improve this effect, by reducing the width of the virtual modules in ϕ, at the expense of z segmentation.

A hardware demonstrator for the tracklet approach has also been developed. The $+z$ half of a ϕ sector has been implemented in a V7-690 FPGA at 240 MHz, without duplicate removal. The latency from DTC input to track output is measured to be 3321 ns. The resource utilisation for a full ϕ sector is 280,000 LUTs, 2722 BRAM and 1818 DSPs. These utilisation and resource numbers do not include duplicate removal.

In comparison to the tracklet approach (as it currently stands), the track finder presented in this thesis has been designed for almost no truncation losses even in the core of dense jets. As a result of this optimisation, this track finding system utilises about 2.4 times more BRAM in comparison to tracklet, but has significantly more margin for high rates and high pileup. The demonstrator results show that both approaches would be able to meet the L1 latency target.

References

1. Ghabrous Larrea C et al (2015) IPbus: a flexible Ethernet-based control system for xTCA hardware. JINST 10:C02019. https://doi.org/10.1088/1748-0221/10/02/C02019
2. CMS Collaboration (2017) The Phase-2 upgrade of the CMS tracker. Technical report CERN-LHCC-2017-009
3. CMS Collaboration (2017) The Phase-2 upgrade of the CMS L1 trigger interim technical design report. Technical report CERN-LHCC-2017-013
4. CMS Collaboration (2015) Technical proposal for the Phase-II upgrade of the CMS detector. Technical report CERN-LHCC-2015-010
5. Xilinx Inc (2011) LogiCORE IP ChipScope Pro Integrated Logic Analyzer (ILA) (v1.04a), DS299. https://www.xilinx.com/support/documentation/ip_documentation/chipscope_ila/v1_04_a/chipscope_ila.pdf
6. Xilinx Inc (2017) 7 series FPGAs data sheet: overview, product specification, DS180 (v2.5)
7. Xilinx Inc (2016) UltraRAM: breakthrough embedded memory integration on UltraScale+ Devices. White Paper: UltraScale+ Devices, Jun 2016, WP477 (v1.0). https://www.xilinx.com/support/documentation/white_papers/wp477-ultraram.pdf
8. ATLAS Collaboration (2013) ATLAS fast tracker technical design report, CERN-LHCC-2013-007
9. Annovi A et al (2017) AM06: the Associative Memory chip for the Fast TracKer in the upgraded ATLAS detector. JINST 12:C04014. https://doi.org/10.1088/1748-0221/12/04/C04013
10. G. Fedi on behalf of the CMS collaboration, Associative Memory pattern matching for the L1 track trigger for the HL-LHC CMS, May 2016, EPJ Web of conferences, vol 127, p 00008. https://doi.org/10.1051/epjconf/201612700008
11. Bartz E et al (2017) FPGA-based tracklet approach to level-1 track finding at CMS for the HL-LHC. In: Proceedings of connecting The Dots/intelligent trackers 2017, Orsay, France. https://doi.org/10.1051/epjconf/201715000016
12. Hoff J et al (2013) Design for a L1 tracking trigger for CMS. JINST 8:C02004. https://doi.org/10.1088/1748-0221/8/02/C02004

Chapter 7
Outlook and Summary

7.1 Demonstrator Scaling

One advantage of a fully time-multiplexed, all-FPGA approach is the inherent flexibility to adapt and evolve the algorithm choices and their implementation. Improvements can come in two forms: changes that fit within the current technological boundaries; and changes that utilise and exploit newly emerging, or available, technologies. So far, the project has benefited greatly from the former, as the exact algorithm and implementation has evolved greatly over the past several years. This section discusses the continuation of these improvements, and the opportunity (and in fact requirement) to take advantage of the latter.

The next step in the development of the demonstrator project is to transition from the Xilinx Virtex-7 FPGA to a modern Xilinx Ultrascale or Ultrascale+ device, with transceivers capable of data transmission speeds of at least 16.3 Gb/s. A system architecture that makes use of this increased link bandwidth would allow the demonstrator to scale to a time-multiplexed factor of 18, the baseline design described in Sect. 3.3. In this design, each TFP board covers a nonant of the detector, for a total of $9 \times 18 = 162$ boards. The number of FPGAs per TFP board is flexible (likely to range from 1 to 3), and will depend on the extent to which the firmware can be optimised, alongside a compromise between latency and cost. The predominant requirement is that the FPGA supports the necessary input band width of about 1 Tb/s, and enough links to take data from all applicable DTCs. Any additional features described in Chaps. 4 and 5 are expected to have a small impact on system resources, in comparison with the firmware optimisations described below.

With the baseline design requiring a time-multiplexed factor of 18, each TFP would have to process the same volume of data at roughly twice the rate of the demonstrator slice described in this thesis. Naively, therefore, one may set an upper bound on the amount of logic needed at twice the requirement of the demonstrator, in order to handle this doubling in processing bandwidth. As shown in Table 7.1, a doubling of the FPGA resources would require a minimum of three KU-115 FPGAs per TFP.

© Springer Nature Switzerland AG 2019
T. O. James, *A Hardware Track-Trigger for CMS*, Springer Theses,
https://doi.org/10.1007/978-3-030-31934-2_7

Table 7.1 The upper limit of FPGA resources required per TFP for the baseline system (with time-multiplexed factor $n = 18$), inclusive of infrastructure logic. The available resources for a number of compatible Xilinx Ultrascale FPGAs are shown for comparison [1]

	LUT[10^3]	DSP	FF[10^3]	BRAM
Upper limit per TFP	1700	16,944	1832	7002
Kintex Ultrascale 115	633	5520	1266	2160
Virtex Ultrascale+ 9P	1182	6840	2364	2160
Virtex Ultrascale+ 11P	1296	9216	2592	2016

On the other hand, the reduction in the number, and increased purity, of candidates out of the HT when using the tilted barrel geometry should mean that up to a 30% reduction in the number of KF workers is feasible without incurring losses due to truncation. It should also be noted that the demonstrator suffers from significant under-utilisation of processing resources, even in the high occupancy conditions of $t\bar{t}$ events with a pileup of 200. As was shown in Sect. 6.4, the HT input is idling around 50% of the time; and the HT output, which was designed to handle the challenging case of a sub-sector containing a jet without efficiency losses due to truncation is usually idling. In addition, even in the centre of very high p_T jets, the state updater block of the KF, which contains the majority of the processing logic of the KF is idling more than 75% of the time. It is expected that by optimising the data-flow throughout the design, taking into account the gains from adapting to the tilted geometry, it should be possible to halve the logic resources required while maintaining performance.

Similarly, while all fabric on the current full chain demonstrator is clocked at 240 MHz, parts of the system have been tested and re-optimised for increased speeds. The HTP router has been built for 480 MHz in the V7-690, and the HTP mathematics block is able to reach 500 MHz. The HT array is capable of running at 480 MHz in the KU-115 (416 MHz for a 32 array per chip implementation), and results indicate that this could improve with continued optimisation. The architecture of the Kalman Filter is such that there are very few fan-outs and the design is heavily pipelined, lending itself towards operation at higher clock frequencies. Operation at 300 MHz has been achieved for the state updater, and higher frequencies still are to be tested. Preliminary studies indicate that the Ultrascale, and particularly the Ultrascale+ FPGAs are much better adapted to running large-scale single clock domain designs across the device, and this can be taken advantage of. Therefore, by targeting a clock speed of 480 MHz for most parts of the system, it should be possible to double the processing bandwidth of the design and halve the resources required per TFP, with respect to Table 7.1.

Considering the combined savings from maximising the processing bandwidth, through both optimising the design to minimise under-utilisation and running the algorithms at higher frequencies, it is highly feasible that a final TFP could be constructed with no more than two KU-115 FPGAs, or one VU-11P FPGA.

In comparison with the TFP demonstrator, the latency of the final track finding system would be reduced when scaled to the baseline design with 16.3 Gb/s links. The data accumulation periods in the KF and HT, where the FPGA must wait for all the data to arrive before it can continue processing, would be reduced by 450 ns each. A conservative estimate of the latency of the baseline system, from input of stub data at the DTC, to input of tracks at the L1 correlator, is given in Table 7.2. The first column shows the latency when all steps of the chain are running at 240 MHz. The second column shows the latency when only the HT and HTP are running at 480 MHz, and the rest at 240 MHz. This scenario has already been achieved in firmware. The third column is more speculative, and shows a scenario where all steps in the chain run at 480 MHz. As the DR and KF firmware have not yet been tested at 480 MHz, an associated latency reduction of 30% is assumed as this is what was achieved with the HTP router when designing for 480 MHz operation. The overall latency of the unpacking, formatting and regional assignment steps in the DTC is conservatively estimated to be 250 ns. An additional latency of <150 ns is incurred for the transfer of data from DTC to TFP. If the entire design could be run faster than 240 MHz, further latency savings could be targeted. By accumulating the second column, one can see that a conservative estimate of the final FIFO latency would be 2.575 μs. If it is possible to run all steps at 480 MHz, an estimate of the maximal FIFO latency can be placed at 2.2 μs.

In parallel with these firmware and algorithm developments, a PCIe processing board, mounted with a KU-115 has been developed at Imperial College London.

Table 7.2 Latency table for the baseline system, extrapolated from the existing TFP demonstrator. One SERDES stage internal to the TFP is also assumed, to cover the expected scenario of two daisy-chained FPGAs. The second column is a scenario in which the HT and HTP run at 480 MHz (as has been demonstrated in firmware). The third column is for a scenario in which all processing steps run at 480 MHz. The latency for the HT and HTP stages at 480 MHz are taken from the firmware described in Chap. 4. As KF and DR firmware have not yet been tested at 480 MHz, an extrapolation to 30% reduction in latency is used in the third column, to account for additional pipelining and clocking registers. This is the reduction in latency that was achieved when modifying the HTP router for 480 MHz operation

Latency [ns]	240 MHz	480 MHz HT/HTP	480 MHz all
DTC	250	250	250
DTC → TFP (SERDES and fibre)	150	150	150
HTP	251	162	162
HT	575	492	492
KF	1220	1220	854
DR	38	38	27
Internal SERDES	120	120	120
TFP → L1 (SERDES and fibre)	150	150	150
TFP First out → Last Out	225	225	225
Total: First in DTC → First in L1	2664	2575	2198
Total: First in DTC → Last in L1	2905	2800	2423

This chip provides fifty-two 16.3 Gb/s bidirectional transceivers. This card will be used to demonstrate the functionality of the proposed 16.3 Gb/s TFP.

7.2 Projected Final System Technology

Taking into account what has been demonstrated, developments are ongoing to determine the specifications of the final DTC and TFP boards.

7.2.1 Outer Tracker Data, Trigger and Control Board

The Outer Tracker DTC must have the ability to talk to the front-end at 2.56 Gb/s in, 5.12 or 10.24 Gb/s out, in addition to the track finder and DAQ at 16 or 25 Gb/s. While it may be possible to build this functionality into a single FPGA (monolithic) design, it would require a very expensive part, and the limited FPGA resources would have to be shared between all firmware tasks. It is therefore preferable to design a board where multiple FPGAs route the data as needed. For example, a system with three smaller FPGAs connected to the front-end, and one larger FPGA connected to the trigger and DAQ could provide plenty of FPGA logic and links, at a more affordable cost. The back-end FPGA could provide 25 Gb/s links to the track finder and DAQ, or a larger number of 16 Gb/s links, at a reduced cost. This will depend on the final design of the TFP and DTH boards. It should be noted that PCBs can be designed with a footprint that is compatible with several different FPGAs (such as the VU-9P and VU-11P). This may allow the board to be designed in advance, and the FPGA to be selected based on requirements at a later date. The same processing board may therefore be used across upgrade projects, with the FPGAs and optics being populated to the needs of the individual user.

7.2.2 Track Finding Processor Board

A balance between cost, FPGA processing power, and bandwidth must be met when specifying the final TFP board. Also in ATCA format, the final board is expected to require two FPGAs, operating in serial (daisy-chain) mode. This is to ensure that sufficient FPGA logic resources are available for the track-finding task, as described in Sect. 7.1.

Figure 7.1 illustrates the potential configuration of FPGAs and I/O that could be used for the final TFP. An important decision will be whether to use FPGAs capable of 16 Gb/s, or 25 Gb/s receivers/transceivers. In a system with time-multiplexing, the minimum value of the time-multiplexing period, P is determined by the time needed to transfer all data from a single event from layer one to layer two. This means

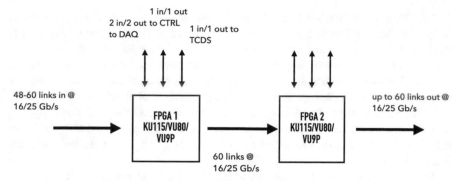

Fig. 7.1 Illustration of the potential FPGA and I/O configuration of the final TFP boards. As a high processing resources to bandwidth ratio is required, it is envisioned to use two FPGAs connected serially. The exact choice of FPGA will be determined by resource requirements, and the decision to use 16 Gb/s or 25 Gb/s optical links.

that in contrast to a $P = 18$ system based on 16 Gb/s links, a 25 Gb/s link system would naturally scale to $P = 12$. Therefore a 25 Gb/s system would reduce the size of the system by one third. It would also lead to a reduction in data-transmission latency. Devices that offer transceiver speeds of 25 Gb/s are, however, much more expensive, and typically have a poor balance between logic and bandwidth (too little logic) when compared to 16 Gb/s alternatives. The number of links per FPGA also significantly impacts the cost. As 48 DTC boards must be connected to each TFP, the chosen FPGA must provide this minimum number of transceivers, in addition to the links to the second FPGA. To be flexible to future algorithm changes, and to have the ability to load balance data streams from busier DTCs, it may be desirable to have more than this amount. Within the constraints of cost, and appropriate logic to bandwidth balance, it is currently only feasible to expect a chip with up to around 120 transceivers [1].

In line with increases in FPGA transceiver speed, optical modules have been developed such as the Samtec FireFly cables [2], which are capable of carrying up to 28 Gb/s over 100 m. As a result of demands from industry, 56 Gb/s and faster active optical modules and FPGA transceivers are becoming available, but FPGAs supporting such speeds are may not be especially useful (or affordable) for CMS, it many lower speed links are required to physically distribute data.

Finally, the choice of number and type of FPGA will also be limited by thermal and power envelopes [3]. While the ATCA standard [4] recommends limiting the front board power consumption to 200 W (with a maximum of 400 W for both front and rear boards occupying a slot), which is significantly higher than the 80 W recommended in the MicroTCA specification, some higher end Ultrascale devices can dissipate above 100 W per part [1]. Xilinx tools report that a KU-115 FPGA with 80% utilisation of all resources, and sixty-four 16 Gb/s transceivers, running at 480 MHz dissipates 68 W. Under the same conditions, (except for a 30% reduced DSP usage), the VU-9P with seventy-two 10 Gb/s and twenty-eight 25 Gb/s active transceivers dissipates

130 W. This larger heat load must be extracted from a card that is only 1.5 times deeper (corresponding to a proportional increase in air-cross-section for cooling) than the MicroTCA form factor [5] used in the demonstrator. In order to do this, the heat sink area and the air speed must be increased. In the simulated examples discussed above, assuming 4.0 m/s airflow and heat sinks of area 10 cm^2 (roughly four times the FPGA package size), FPGA temperatures of 59 (KU-115) and 62 °C (VU-9P) are predicted. In addition, it is worth noting that the high speed optical modules are particularly sensitive to temperature, and typically must be operated at less than 50 °C if they are expected to last for around 15 years with less than 1% failure [6, 7].

Within these constraints, however, there is still a large degree of flexibility to develop track finding algorithms, and future work will involve optimisation of the algorithms that will run on these chips, with a goal to maximise performance based on the requirements of the L1 trigger. This may involve modifications to the firmware, such as: improving 2 GeV track finding, electron efficiency, or displaced vertex tracking; while minimising cost and latency wherever possible.

7.3 Summary

A hardware demonstrator has been assembled in order to prove the feasibility of a track finder at L1 for CMS at the High Luminosity LHC. The demonstrator implements a Hough transform algorithm for coarsely identifying track candidates and a Kalman filter to clean and fit them, on FPGA-based hardware, along with corresponding emulation software. The hardware demonstrator has successfully shown that track finding and fitting for charged particles with transverse momentum exceeding 3 GeV is possible at 40 MHz, and within 4 μs, in the challenging high occupancy conditions of the HL-LHC. This has been accomplished using currently in-hand technology (MP7 processing boards), and one can expect the latency and projected scale of the system to be reduced as algorithms are optimised and refined, and new technology becomes available.

References

1. Xilinx Inc (2017) UltraScale architecture and product data sheet: overview, product specification, DS890 (v2.11). https://www.xilinx.com/support/documentation/data_sheets/ds890-ultrascale-overview.pdf
2. Samtec Inc. (2016) FireFly optical half cables application note, revision 1. http://suddendocs.samtec.com/notesandwhitepapers/firefly_half_cable_app_note.pdf
3. Bobillier V et al (2016) MicroTCA and AdvancedTCA equipment evaluation and developments for LHC experiments. JINST 11:C02022. https://doi.org/10.1088/1748-0221/11/02/C02022,http://iopscience.iop.org/article/10.1088/1748-0221/11/02/C02022/pdf

4. PICMG (2003) AdvancedTCA short form specification. https://indico.cern.ch/event/119030/attachments/61294/88092/PICMG_3_0_Shortform.pdf
5. PICMG (2006) Micro telecommunications computing architecture short form specification. https://www.picmg.org/wp-content/uploads/MicroTCA_Short_Form_Sept_2006.pdf
6. Private correspondence, Rose A. Electronic Engineer, Department of Physics, Imperial College London, andrew.william.rose@cern.ch
7. Private correspondence, Iles G. Electronic Engineer, Department of Physics, Imperial College London, gregory.iles@cern.ch